A PHILOSOPHY OF FLIGHT

by

John C. McCollister, Ph.D.

JSB ENTERPRISES

To: John McCollister
With respect,
John McCollister

Library of Congress Cataloging in Publication Data:

McCollister, John.
A philosophy of flight.

1. Flight — Moral and religious aspects. I. Title.
TL553.M4 629.13'01 80-83146
ISBN 0-934086-12-5 AACR2

To "Rab"

Cover and illustrations by Kathleen M. Fosbinder

ISBN# 0-934086-12-5

Library of Congress # 80-83146

JSB Enterprises
763 Pinetree Drive
Indian Harbour Beach, FL 32937

TABLE OF CONTENTS

Table of Contents (cont'd)

Introduction

It is too much to expect that the average person, quite naturally wanting to get from Point A to Point B as safely, quickly and comfortably as possible, should look on flight as something akin to a personal religion. Yet it is exactly that to more men and women than is commonly imagined.

To some, it is a way of life involving deep and intense commitment, both professional and emotional.

To others, it is an expression of freedom—a way to "slip the surly bonds of earth," as pilot/poet John Magee wrote in his famous poem, *High Flight*.

And to still others, it is indeed a feeling that flying mysteriously brings one closer to God.

Perhaps this explains why so many great books about aviation have been written by men who were simultaneously pilots and philosophers. Antoine de Saint Exupery, Ernest Gann, Nevil Shute and David Beatty—were or still are airmen who sensed the mystique of flight, and capable of translating these feelings into mirror images of what lay in their own hearts and minds.

And why not? To the flier, provincial and nationalistic boundaries are just artificial lines on an airway map. It always has seemed significant to me that H.G.Wells, in his futuristic novel *The Shape of Things to Come,* decided that Utopian peace would eventually be achieved—not by politicians, statesmen and generals—but by airmen who simply extended their own brotherhood of the skies into a vaster brotherhood of man, because they could envision with terrible clarity the only alternative.

This is why I regard John McCollister's new book as an important contribution to aviation literature. Certainly no man is more eminently qualified to author such a work. He is a pilot, an ordained Lutheran minister, an articulate speaker and lecturer, and a man of humility and humor—a priceless combination.

Nor can I think of a more opportune time for a book like *A Philosophy of Flight*. All aviation is going through a difficult period, for in these days of an anti-technological environment, aviation's very status as the epitome of technological achievement makes it a most vulnerable target.

But it is far more than technology John McCollister so skillfully and brilliantly expounds in these pages. As such, it should be required reading for those who love flight, and also for those who do not really understand it.

—Robert J. Serling

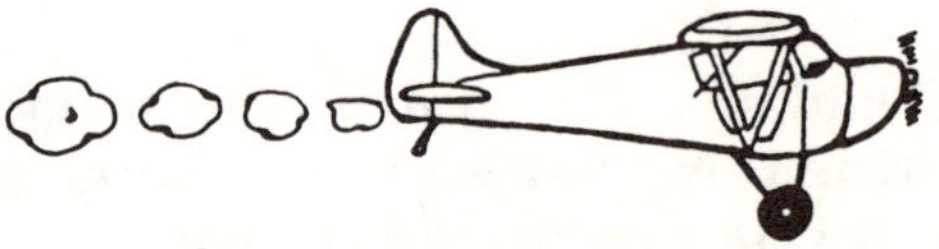

''In the beginning, man created the airplane...''

Birth

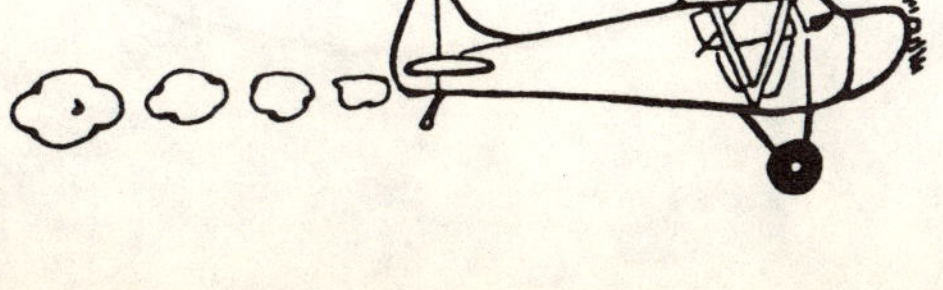

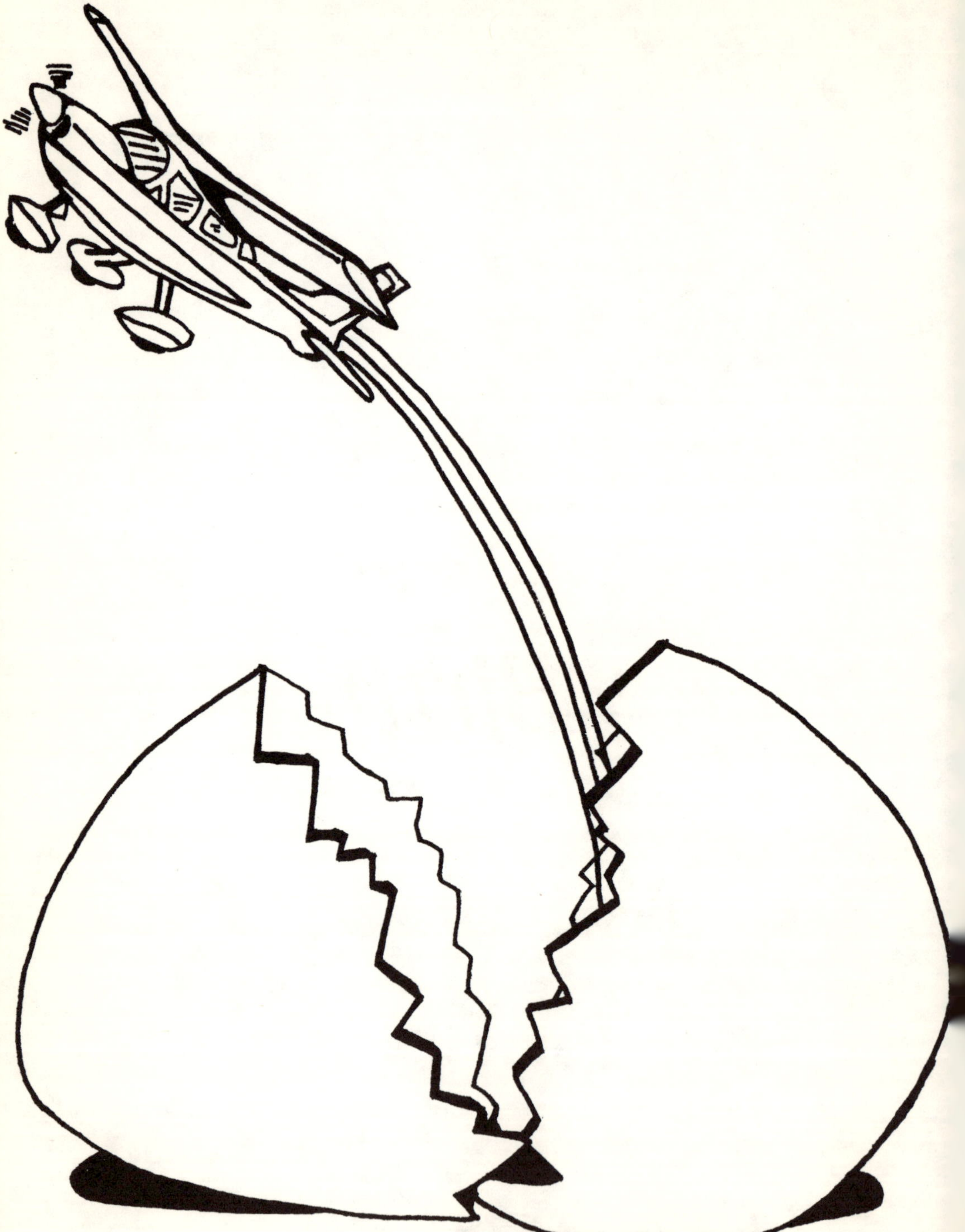

GENESIS

In the beginning, man created the airplane.

And man said: "Let there be a *propeller* to bite into the wind. Let it pull smoothly, steadily, turning at an incredible rate of 40 to 50 times per second."

And man said: "Let there be *wings* to keep it afloat on an invisible sea of air, controlling the forces of gravity through laws of physics, calculus, and aerodynamics."

And man said: "Let there be *controls* to maneuver each creation in a variety of configurations from straight and level flight to snap rolls and outside loops."

And man said: "Let there be *radios* to maintain communication with those earthbound souls who advise and control the highways to adventure."

And man said: "Let there be *instruments* to guide through clouds, fog, rain, or snow. Safely. Surely."

And man said: "Let there be *airports* in a variety of shapes and sizes scattered throughout the land so that wherever it may fly, the airplane always will have a place it can call 'home'."

Finally, man said: "Let there be *pilots* to have dominion over the rest of this creation. Urge them never to abuse it, nor to exceed its limitations, but to care for it, to pamper it, and to respect it."

After this, man beheld his creation. And, lo, it was good. Very good, indeed.

MAN IN FLIGHT

From the earliest days of creation, it was a mere dream — nothing else. For centuries, philosophers thought about it, journalists wrote about it, artists painted pictures about it, and scientists formed theories about it. But it remained a dream... just a dream.

Then, one December morning on a hill in North Carolina, something happened to change all that. Two unlikely brothers — owners of a Dayton, Ohio bicycle shop — transformed that dream into a reality with a flight of twelve seconds covering 120 feet.

From Kitty Hawk to Cape Canaveral, the story of flight has shaped our lives like none other. Its pioneers — Lindbergh, Earhart, Glenn, Armstrong — are much more than heroes; they're legends who approach royalty. Certainly, our lives would not be as rich without them.

And it all began with a dream — a dream bigger than life itself.

Were our dreams limited to what others said could not be done, or to what we ''knew'' to be unattainable, would we have been able, ever, to leave the ground?

In a sense, the greatness of any human being is in direct proportion to the dimension of his dreams.

The pilot, by his very nature, is willing to dream beyond the obvious.

''Happy are those who dream dreams
and are ready to pay the price
to make them come true.''

—Cardinal Suenens

NOVELTY
INFANT
NOVELTY
NOVELTY
NOVELTY

INFANTS

On December 17, 1903, when two bicycle makers from Dayton, Ohio powered a biplane through the air for twelve full seconds, only a handful of newspapers reported what must be regarded as one of history's most important happenings. Reporters seemed more interested in covering the more interesting news about Christmas shoppers and parties hosted for New York debutants.

A teenager in Germany, Wernher Von Braun, experimented with a new-fangled device called a "rocket." Exciting, perhaps, but few took his work seriously.

Similarly, when Michael Faraday invented the electric generator in 1831, a fellow scientist scoffed at the invention by shrugging his shoulders and asking: "What good is it?" Undaunted, Faraday responded with another question: "What good is an infant?"

Michael Faraday did not predict city streets lit as brightly at night as at high noon. The Wright Brothers did not envision a 747 leaving jet streams in the sky. And Wernher Von Braun merely fantasized about a man setting foot on the moon. At the same time each could see tremendous potential within the infant stages, while others gave them little more than passing notice.

During the summer of 1980, a ninety-pound woman crawled into a contraption resembling bamboo sticks separated by sheets of "Saran Wrap," and flew non-stop for one mile. What was so amazing is that the "engine" was powered only by the sun.

Because of a looming energy crisis, the media reported the event. Yet most commuters viewed this as a novelty—a change of pace in the humdrum news of the day.

Again, the hard-nosed skeptic asked: "So what?"

A novelty or an infant? To discern the difference is the mark of genius.

SOUL
OH!

SOUL-OH

Few of us take up flying in order to realize higher states of awareness. It is quite enough to get the plane in the air, maintain a safe altitude, and set it down again. But, alas, there *is* something extrasensory, superhuman, that happens to the man who flies his first solo. Some call it a religious experience.

He's studied all the holy manuals, and bowed to the hangar messiahs. He's worshipped the famous pilot-idols, genuflexed when receiving their directives, crouched under their wings, and followed their guidance and dogma with blind faith until the first solo.

When the great day comes, he's, now, on his own, just as God created him. It is *now* — not past, not future — when a state of "instant being" overtakes him. It is now time to break the cocoon of innocence and discard the pious yoke of the disciple. Now is the time for revelation in the truest sense of the word.

He latches the cockpit door, completes the check list, receives the necessary clearances, points the plane down the runway, pulls back on the stick, and rises to the heavens above useless wheels.

Time for prayer is over. Time for following like a hypnotized sheep is done. He gets, as Socrates advised, to know himself. His new being rides like the emerged butterfly — on the cosmic breath of the All-knowing.

Up, up he climbs. Soaring. Turning. Growing in first-hand knowledge of creation.

And when it's time to refuel, he can alight on the earth, never again to be its captive, grounded in ignorance.

Fear and doubt voluntarily have been crucified on the cross of experience.

He has risen. And greater things will he do.

SQUARE
ONE

BEGINNING AGAIN

For those of us who lived through World War II, the date of December 7, 1941 holds a special meaning. It marked a drastic interruption of a way of life — a surprise attack...a radio announcement...a nation in shock...a Presidential declaration....

Driven back to its shores and bombed into final submission, this nation, standing on the brink of annihilation, was disgraced before the world.

Yet, in the midst of pending doom, this tiny nation found a place of beginning again. It restructured its government, raised its standard of living to its highest level in history, and successfully manufactured new products for export. In fact, newspapers recently reported Japan's noteworthy contributions to the general aviation and business jet markets.

Perhaps some of us still feel the stings of hostility resulting from the attack on Pearl Harbor. Nonetheless, as exemplified by its accomplishments in the aviation industry, we must admit that Japan showed what can happen when someone chooses to begin again.

Individuals, as well as nations, find themselves needing a place of beginning again.

The struggling artist, disappointed with his latest painting, approaches the canvas with new hope in his heart. The writer drops his seventh attempt into a well-filled wastebasket, and inserts another sheet of paper into the typewriter. A student pilot who has just failed his F.A.A. written examination, dusts off his textbook, and reads for a clearer understanding.

In a certain sense, everyone of us, every day, needs a place of beginning again. The opportunity is there; it's up to us to search it out.

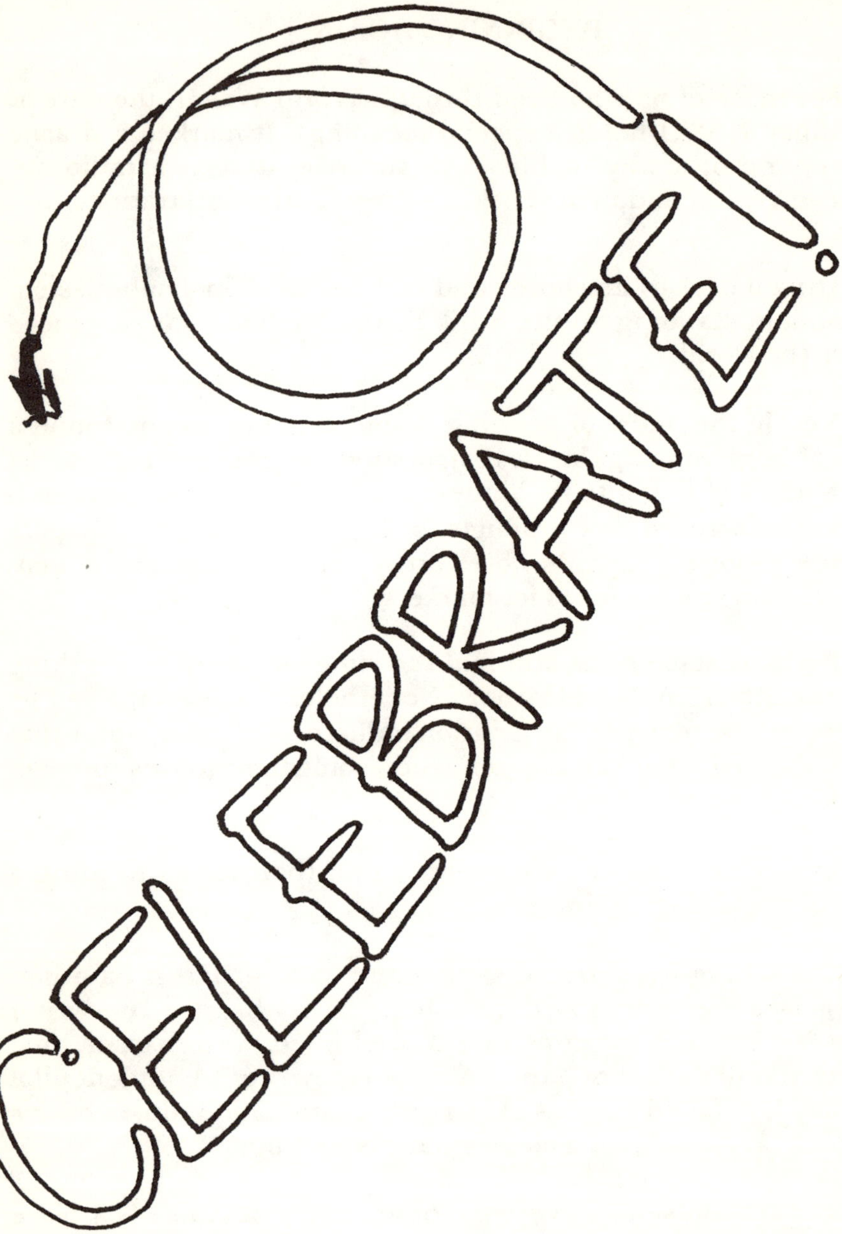
CELEBRATE!

CELEBRATE

A scientist once calculated that if the time from creation until the present were squeezed into one year, man would appear in the last ten minutes and recorded history in the last ten seconds.

Yet within that relatively brief period of time, look at what man has accomplished, especially in aviation technology. From a hop of 120 feet — less than the length of a 747 — to a successful landing on the moon, all within a microscopic span of 75 years. And it began with that daring twelve-second journey at 10:35 on the morning of December 17, 1903 at Kitty Hawk, North Carolina by two bicycle builders from Dayton, Ohio.

Each year, as we approach the anniversary of man's first flight, the question is asked: "How can we commemorate this historic milestone?"

On each December 17th, it seems, a youngster builds a scale replica of Orville's and Wilbur's original flying machine, a group visits the monument at Kitty Hawk, and someone attempts his own 120 foot flight at a local airport. Each, in his own way, marks the first successful powered flight.

But of what genuine value is the building of a model airplane, or a pilgrimage, or for that matter, another 120 foot flight? I suggest that we best commemorate this anniversary through something more substantial.

I propose that on December 17th, at 10:35 a.m., wherever we may be, each of us stop what we are doing, and offer a pledge of rededication to *precision* and the sharpening of our skills as pilots, engineers, and mechanics...to *adventure* and the conquering of new worlds...to *safety* and the growing awareness of our responsibility to our aircrafts, to others, and to ourselves.

By so honoring the Wright brothers' success, we pay homage to the whole of general aviation. To recall man's first flight through a one-shot event is not justice. For what remains after the project is completed, the candles extinguished, and the last drop of champagne consumed?

Better that we dedicate ourselves to precision, to adventure, and to safety. In that way, the party continues all year long — from one anniversary to the next.

Celebrate!

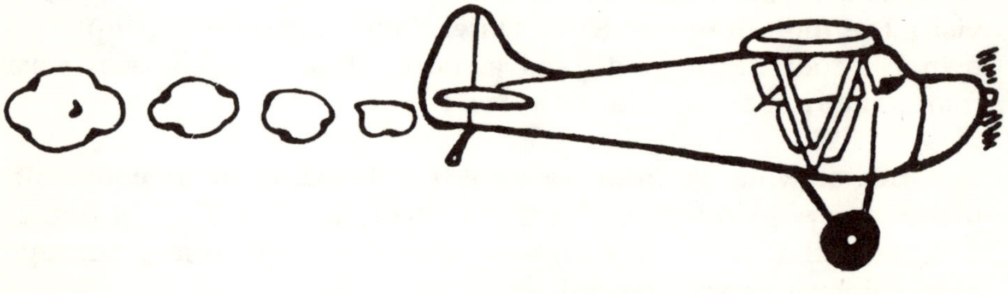

''If you know what to listen for, it's another kind of music.''

Life

ROAR

A POINT OF VIEW

"How can you stand the noise?"

This is the most frequent question I hear from friends who visit my office on the campus of Embry-Riddle Aeronautical University located at the busy Daytona Beach Airport. Sometimes I even get this question from someone on the other end of a telephone conversation as an Eastern Air Lines 727 roars off Runway 6, climbing directly over my office building.

My answer: "I love it."

Don't get me wrong, please. I am as concerned as the next person about noise pollution. But airplane sounds are different. It depends upon your point of view.

First, there are the constant sounds of single-engine aircraft taking off and landing. Many contain students who are experiencing the unforgettable thrills of a first flight...a first solo...a successful check ride....

Then there are the sounds of the twins and corporate aviation jets logging business flight hours — a commentary on the right of every citizen of the United States to engage in free enterprise.

About those sounds of the big guys. They may carry a young, enthusiastic salesman about to land an important account; or a weary executive and his family as they break away from the hustle and bustle of New York City for the quiet retreat of a Florida beach; or two starry-eyed honeymooners who, just a few hours before, pledged to love each other for better...for worse; or a grandson on his way to attend the funeral of one whom he has loved throughout his life.

Each sound, then, has a unique story — never told before; never to be told again.

Noise? Not really. If you know what to listen for, it's another kind of music.

WORDS

Words can be a jungle, or words can be a garden of delight.

Words can be wild weeds that ravage the soil and rob its nutrients, or they can yield a harvest of beauty, and comfort, and consolation, and hope.

Words sloppily selected and thoughtlessly spoken can be a source of grief and contention. Words well-pruned and carefully chosen can be a balm to the oppressed and distressed.

For the private pilot, certain words hold significant meanings. For him, the following are among the more expressive:

The most peaceful are VISIBILITY UNLIMITED.

The most adventuresome are CLEARED FOR TAKEOFF.

The most tragic are THERE HAS BEEN AN ACCIDENT.

The coldest are YOU ARE GROUNDED.

The most comforting are I HAVE YOU IN SIGHT.

The cruelest are THIS AIRPORT IS CLOSED TO PRIVATE AIRCRAFT.

The most regretted are I DON'T HAVE TIME TO WAIT FOR THE WEATHER TO CLEAR.

The most hopeful are YOUR LANDINGS ARE MUCH SMOOTHER.

The most welcomed are YOU PASSED YOUR FLIGHT EXAM.

The most inviting are COME FLY WITH ME.

The most reverent are THE SKY.

Words. To the pilot, they are special. The wrong word at the wrong time is a piercing arrow; the right word at the right time is a salve.

TIES AND AIRPLANES

It was bound to happen sooner or later.

It hit me, yesterday, when I took my semi-annual pilgrimmage to the local men's clothing store. There, as big as life, was a rack of new ties for the fall season. You didn't have to be Oleg Cassini to realize that the ties were much narrower than the one I was wearing at the time.

The man behind the counter told me the bad news. It seems that the tie, so popular in the 1950's, is staging a comeback. "Wide ties are 'out,' " he said with a hint of smugness.

"But I haven't worn some of the ties I got last Christmas," I pleaded.

"Sorry," he said. "I've got no control over these things. They have decided that this is the shape of ties to come."

I've yet to discover who "they" are who dictate such trends. I envision some mysterious committee meeting in a remote "never-never land" taking votes as to what I will or will not wear this year.

I suppose I could launch a one-man campaign against this change, but what good would that do? I'd be outnumbered. Perhaps ostracized. My friends would sneer at me behind my back. My boss might conclude that I was not the progressive employee he had pictured.

So, bending to the imposed standards, I purchased two new ties.

But that's fashion for you. One year it's wide lapels; then narrow. Ladies' hem lines have never been consistent.

Ah, but airplanes are different. Vastly different. What type ever goes out of style? The man with a Waco biplane is just as accepted in the aviation community as the one with a Learjet.

Neither apology nor alibi is needed for either. Each type of aircraft promises a unique blend of adventure.

I guess that's just another of the beautiful sides of aviation.

In the meantime, anybody wanna buy a wide tie...cheap?

CONTROLLED ROUGHNESS

Richard Bach once told this writer that, on several occasions, he tried to rewrite his classic: JONATHAN LIVINGSTON SEAGULL. "I want to smooth out some of its rough edges," he said, "but everytime I attempt to do so, it never comes out better."

Yet, perhaps, this is what makes JONATHAN live. The story has a "controlled roughness." It's not plastic, or too smooth, or antiseptic. In fact, we see in the story of JONATHAN those things we overlook in ourselves — the imperfections that create the intangible dimension called: "personality."

Those rough edges in our make-up — you know, the ones we struggle to change — a funny mole on a chin...a receding hairline...an odd shaped nose.... Why not accept these things, control them, and build upon them?

Woe be unto us were everything completely smooth. Why, there would be no hills to climb or hurdles to overcome. We would become soft, weak, and bored to tears. In fact, there would be no airplanes to fly. It is the "controlled roughness" of the stabilizer, the rudder, or the ailerons that lifts us off the ground, and produces a standard-rate turn to a new path of freedom.

Roughness — by itself may be considered unsightly, homely, or evil. But "controlled roughness," especially to the pilot, can be the most beautiful thing in the world.

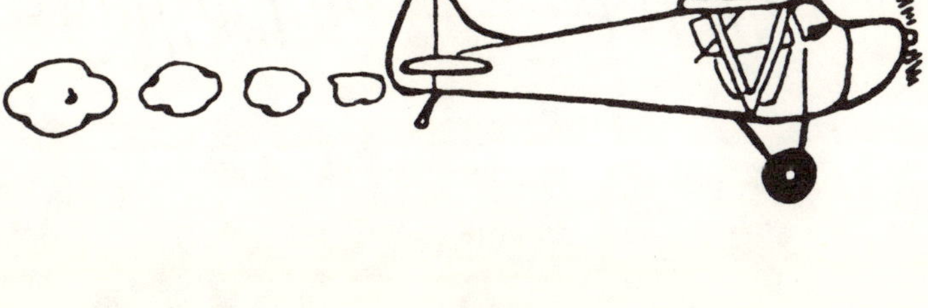

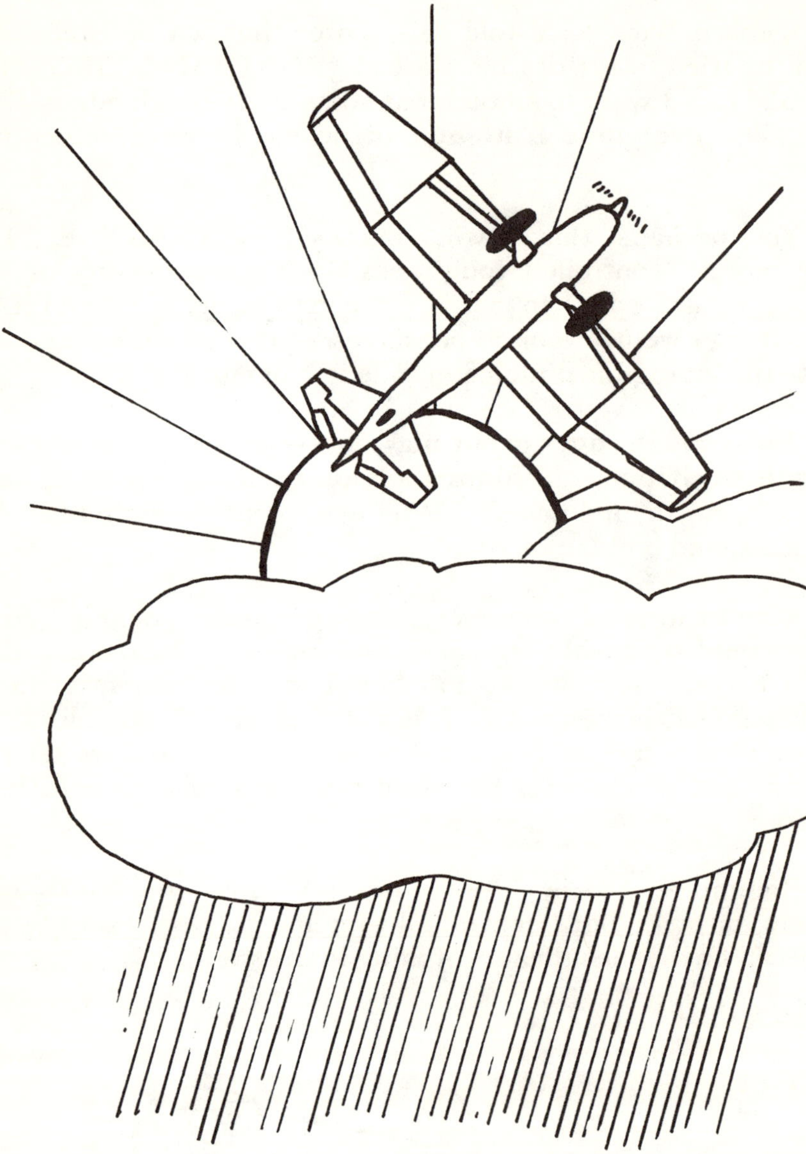

CONTROLLED ROUGHNESS

Richard Bach once told this writer that, on several occasions, he tried to rewrite his classic: JONATHAN LIVINGSTON SEAGULL. ''I want to smooth out some of its rough edges,'' he said, ''but everytime I attempt to do so, it never comes out better.''

Yet, perhaps, this is what makes JONATHAN live. The story has a ''controlled roughness.'' It's not plastic, or too smooth, or antiseptic. In fact, we see in the story of JONATHAN those things we overlook in ourselves — the imperfections that create the intangible dimension called: ''personality.''

Those rough edges in our make-up — you know, the ones we struggle to change — a funny mole on a chin...a receding hairline...an odd shaped nose.... Why not accept these things, control them, and build upon them?

Woe be unto us were everything completely smooth. Why, there would be no hills to climb or hurdles to overcome. We would become soft, weak, and bored to tears. In fact, there would be no airplanes to fly. It is the ''controlled roughness'' of the stabilizer, the rudder, or the ailerons that lifts us off the ground, and produces a standard-rate turn to a new path of freedom.

Roughness — by itself may be considered unsightly, homely, or evil. But ''controlled roughness,'' especially to the pilot, can be the most beautiful thing in the world.

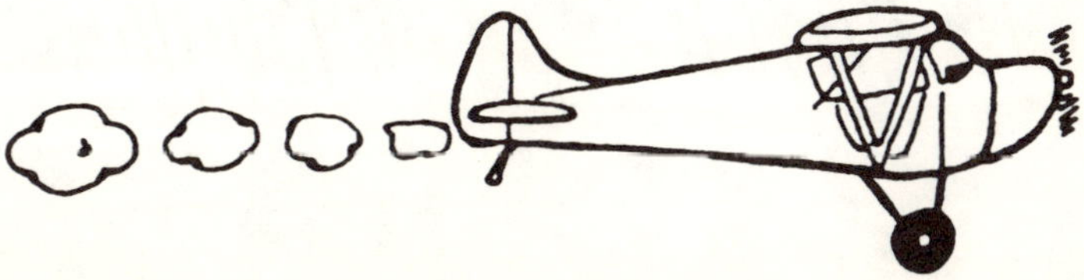

STAYING ABOVE IT ALL

Picture, for a moment, a small bug that has wandered into a plush mansion. It makes its way into the living room, and crawls onto a multi-colored Persian rug. Suddenly, it finds itself amid a kaleidoscope of brilliant hues. As the bug travels from one area of the rug to another, its mood and feelings are affected by each particular color.

The bug first enters a red section. The surrounding crimson atmosphere makes it angry, uptight, ready for a fight. It crawls into a blue area, and becomes calm and peaceful. In yellow, it feels cowardly. In white, unclean and unworthy. Back to red, and it's angry again.

It looks like our bug is headed for a nervous breakdown!

Poor little bug. If only it could benefit from another viewpoint — that of being *above it all*. Then, up there, it could see how the many colors weave themselves into a magnificent, harmonious pattern.

A pilot has this advantage. Whenever he feels pulled down by the mundane problems of life, he can head for the nearest airport, jump into the left seat of his favorite airplane, and soar into the heavens above the earth's surface.

Below him, the world appears different. He notices an overall pattern of which his problems are but a small part. It's not that he avoids his problems, but, for a while at least, he puts them into proper perspective.

After he lands, he is better equipped to deal with them and, perhaps, better able to solve them.

Now, what's that problem that's been "bugging" you?

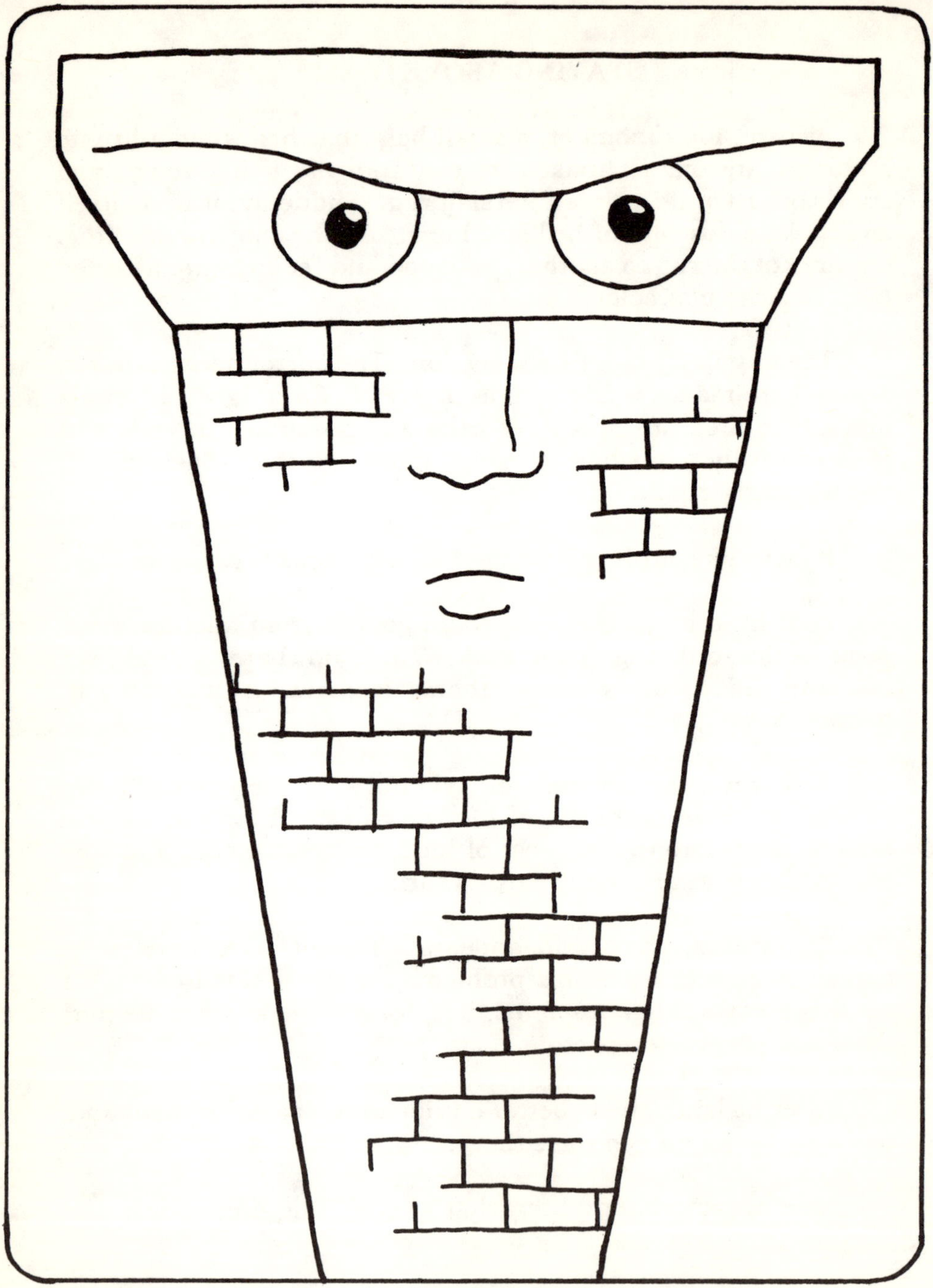

TOWERVISION

It was a beautiful Sunday afternoon in early April when I flew a Rockwell Commander 112 from Daytona Beach, Florida to Washington, D.C. The weather was CAVU all the way. As I approached National Airport, I was happy to see no other aircraft for miles around. Since my fuel supply was diminishing, and I was tired from the nearly five hour flight, I wanted to land as soon as possible; I quickly called the tower operator for instructions.

"Enter a long downwind pattern," he advised in a rather carefree manner. "You'll be number two to land."

"Number two?" I asked myself. "Why won't he give me a quick, straight-in approach?"

I picked up the microphone in order to debate the issue when, suddenly, my rebellious ego was catapulted back to reality. Coming out of nowhere was the largest 727 ever manufactured (at least, it seemed that big to me). Wow! Had I attempted to land when *I* wanted to, my single-engine airplane would be no match bumping heads with that 727.

I guess that's why the tower operator is the boss. From his position, along with the help of radar and other navigational aids, he can see the entire control zone.

Yet, isn't this the same sort of perspective that God has as compared to us?

Here on earth, we make our demands, recite prayers, and plan schedules on the basis of a limited point of view. When we don't get what we want, we grumble, and shake clenched fists to the heavens. However, like the tower operator, the Lord can see the whole picture — not only in terms of right now, but also concerning the events of tomorrow...next week...next year....

Therefore, when I read about what God says as to what I should do or how I should act, doesn't it make sense to pay close attention and follow directions?

THE MANUAL

At last, the big day is here. Ever since the "bug" bit you, you've saved and sacrificed for it. You finally purchased an airplane.

After signing the necessary papers, you cross the dealer's parking ramp, climb into the cockpit, start the engine with just one turn of the key, and taxi the plane to the runway.

You have a rewarding sense of pride as you listen to the smooth purring of the engine. With a mere touch of the brakes, you make the airplane stop in its tracks.

After running through the check list, you move onto the runway, apply full throttle, and feel the surge of power pulling the airplane faster...faster...higher.... It's beautiful!

During the flight, you and the plane get to know each other better. It's a love affair, and the two of you become like one.

You return to the airport, and park in front of your newly rented hangar. The engine is turned off. In the silence for a few moments, you sit appreciatively absorbing the sensation of utter satisfaction.

Then, with a bit of curiosity, you pull from the map pocket an operating manual prepared by the manufacturer. As you leaf through it, you find some sound rules for the proper care of your airplane — which octane of fuel to use, a certain air pressure in your tires, suggested times for oil change, these and many others, each of them important.

But you say to yourself: "This is my airplane. I can do with it as I please. I can use a cheaper gasoline and save money. I'll carry whatever air pressure I like in my tires. I'll just add oil, not change it. No one is going to tell me how to run my own airplane."

That's your privilege, I suppose. It's your airplane, all right. Yet let us assume you do as you please. You may keep your independence and self-esteem, but soon that magnificent airplane will become a mechanical cripple. What a shame. All because you ignored the directives of the manufacturer.

Nearly 4,000 years ago, mankind was given a manual of operations, too. It contained only ten short items called "Commandments." If you haven't looked at them lately, it might be good advice to review them, and try your best to follow these directives of the manufacturer.

REV. PILOT

A private pilot is a preacher of sorts.

Long ago, he was baptized in the name of flight and resurrected from mere existence on the ground to be born again living in the sky. Now, he's an evangelist, bent on converting any non-believer.

At a party, at work, on the golf course, he preaches the good news. His congregation consists of whoever will listen.

His sermons dwell upon subjects ranging from his personal testimony: "How I discovered flying" to the altar call: "Come fly with me."

He borrows from the scriptures of positive experience beginning with a wistful gaze at the sky, through his first solo, to the final check ride for his license, all the while, eschewing the horror stories of near misses and emergency landings for the sheer adventure and romance of the air above.

His gospel is that, with few exceptions, anyone can fly, and they miss a lot who refuse to try at least once.

Oft times, he confesses his sins of failure to abide by the commandments of safety, but he vows to amend his ways, conscious of the truth that disobedience seldom carries with it forgiveness.

He willingly supports his religion with tithes for flying lessons, fuel, rental fees, or payments on an airplane all his own.

And, each night, before he falls asleep, he offers a silent prayer of thanks for his blessings as a member of this chosen people called "pilots."

Then I discovered FLYING...

YAHOO!

GETTING HIGH

Every year at World Series time, I recall the October Classic of 1956 highlighted by the perfect, no-hit, no-run game that the New York Yankees' Don Larsen pitched against the Brooklyn Dodgers. I'll never forget the reaction to the last called strike on pinch hitter Dale Mitchell. Catcher Yogi Berra ran from behind home plate and jumped into Larsen's arms. Larsen and his teammates — Mantle, McDougald, and Martin — joined in leaping as high as they could in a crazy, uncontrolled celebration to this never-before-shared happening. Many of the 64,419 fans in Yankee Stadium poured onto the field, also jumping for joy.

Today, while watching a video replay of their antics, some dispassionate soul may shake his head in amazement. "Is this any way for civilized people to act?" he asks.

Why not?

A prima ballerina leaps high on stage to the thunderous applause of her audience.

A child screams and jumps erratically at the sight of gifts under a Christmas tree.

Leaping from the earth and reaching for the sky is only natural at moments such as these.

There is one problem, however. In each instance, the ecstatic being returns unceremoniously to earth within seconds due to the universal force of gravity. The moment of exultation is just that — a moment.

Perhaps this is just one of the secrets of the lure of flight, known only to those who dance with the clouds. Unlike the athlete, the dancer, or the child on Christmas morning, the pilot does not have to return to the confines of earth immediately. He lingers and drinks in the glories of absolute freedom apart from the ground. His leap into the heavens is measured not in seconds, but in hours.

Would he trade this opportunity even for a World Series no-hitter?

THE FUTURE OF AVIATION

PREPARING FOR THE UNEXPECTED

One of the thrills I enjoy while teaching at Embry-Riddle Aeronautical University comes with my daily contacts with energetic young men and women who are learning a skill in some dimension of aviation. These wide-eyed enthusiasts spend most of their waking hours preparing to become aeronautical engineers, airport managers, mechanics, pilots, or airline executives. Armed with a unique array of talents, each one strides toward a worthwhile, predetermined goal.

Yet, I wonder just how many of them will be doing the specific work for which they are preparing, let's say, ten years from now. Oh, it's not that the students lack the skills or intelligence necessary for success, but aviation, itself, is never static. Its needs and demands differ from what they were only a few years ago.

We don't have to flip the calendar back very far to the era when aircraft design, for example, used to be a "hit and miss" proposition; today, it is a sophisticated science with drawing boards containing pencilled designs aimed at years in the future and wind tunnels to measure the results.

We no longer dare call mechanics "grease monkeys." They are certified technicians who must know the "ins" and "outs" of complex engines and streamlined airframes.

Electronic experts have revolutionized avionics and altered navigation procedures.

Even airline attendants can be married now.

This constant change should come as no surprise. The very nature of aviation compels us not to gaze in awesome retrospect to the "good old days," but to search for the adventures of tomorrow as the truly glorious times.

Periodically, then, I remind my students that ten years from now, half of what they have learned will be obsolete. The only trouble I have at the moment is that I don't know which half that is.

SUPERPILOT

Quite recently, movie theaters were showing a popular film — a reincarnation of the familiar comic strip character, Superman.

We know the story line: A young lad is sent from the planet Krypton to Earth; he is raised by a humble Midwest family; assumes the disguise as Clark Kent, mild-mannered reporter; carries on a platonic relationship with one Lois Lane; and fights crime through superior strength, X-ray vision, and an amazing ability to fly.

Once you saw the movie, you were tempted to gaze heavenward and wish to yourself: "If only I could fly like that..."

But, think about it for a moment. What does Superman have over us who fly airplanes: Does he really have an advantage? Hummmm. Let's see.

In the first place, Superman cannot fly without wearing that stupid costume. Where can you go dressed like that? The private pilot doesn't change clothes before going to the airport. Besides, there aren't that many telephone booths around.

For some reason I've never understood, Superman is compelled to hide his identity. What a horrible frustration knowing the joys of flying without being able to talk about them. The private pilot enjoys not only the thrill of flight, but also the pleasure of reliving it everytime he talks about it.

Superman can fly only by himself. Who else could keep up with him? The private pilot, on the other hand, would never leave his girlfriend behind. He can take her on a Sunday picnic 100 miles away, or share the fellowship of his comrades at a fly-in at a neighboring state.

Superman travels "faster than a speeding bullet." Now, is that any way to see the sights? To the private pilot, the scenes below tell fascinating stories; they're not lost in one, big blur.

Superman has no real challenge. Each flight for him is perfect, with a methodical, boring sameness. Do you ever remember him correcting for a wind factor? Or consider a magnetic deviation? Or take a weight and balance check? A private pilot gains an unmatchable satisfaction from recognizing these challenges and working in harmony with them.

Finally, Superman is without that one necessary ingredient that everyone must have sooner or later — an airplane to love and pamper. Imagine the emptiness he must feel knowing that he will never know such love.

Fly like Superman? No thank you. After all, why should I miss the fun of flying?

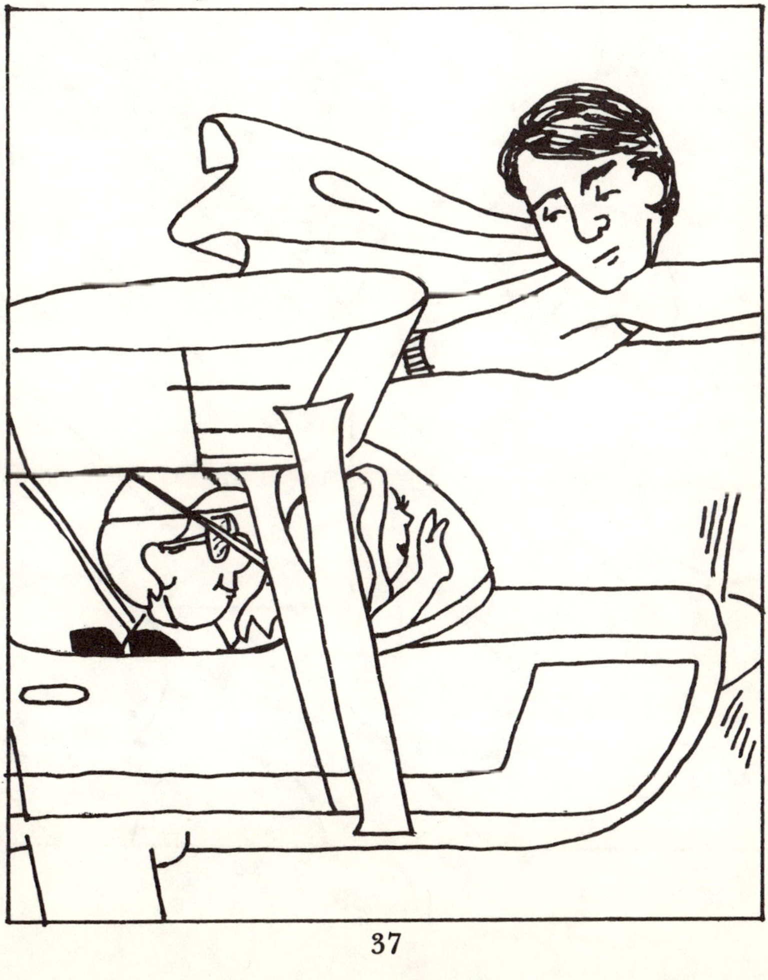

GROUNDED

Have you ever thought about how much we value winning in our society? A famous football coach once claimed: "Winning isn't everything; it's the only thing." Nobody wants to lose.

Realistically, however, this can't be possible. We experience the pain of loss every day of our lives. We lose time, we lose pens and pencils, nickels and dimes. Sometimes we lose our patience or our sense of humor.

Some losses are fairly easy to get over; others cost us a great deal of grief.

To a pilot, losing his privilege to fly — no matter how briefly — is hell. Whether the reason is due to medical or legal reasons, the pilot is frustrated as he remains confined to the ground while his friends enjoy the freedom of the sky. In his frustration he shakes his fist at his Creator or the FAA (whichever is closer). Be it man or God, nobody is going to tell him what to do.

But a lot of good can come from such loss.

You see, when things are all going our way, we tend to know it all. We don't need rules...other people...the FAA... Being grounded — even temporarily — gives us time for reflective thinking and can open the door to a new attitude toward those people and organizations dedicated to insure our safety.

Grounded? To some this spells: "Defeat." To others, this is an opportunity for growth. They find peace despite the pain, calm despite the turmoil, gain despite the loss.

If this sounds strange to you, that's because it IS strange. It's the strange and mysterious way by which we remain "normal."

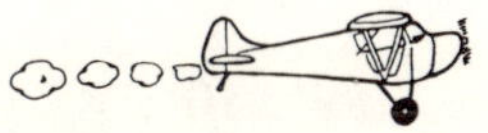

SCHENIC
MOUNTAIN

EDUCATION

Ever since I was a little boy, I've been taught that an educated man is one who reads books. I suppose we're all brainwashed into thinking the same thing. Notice, if you will, how Madison Avenue advertising executives are quick to pick up on this theme.

A recent magazine pictures a bottle of whiskey resting alongside a leather-bound volume containing the complete works of Shakespeare, implying that intelligent people drink the stuff. A man or woman appears scholarly when photographed in front of library shelves lined with books authored by anyone from Dickens to Spillane.

Books and education. They naturally go together.

But books leave something to be desired. They merely report about people, places, and things. At best, they offer second-hand testimony. Descriptions of events are filtered through the perceptions and biases of the writers.

How much more rewarding to be somewhere and, for yourself, witness an important happening, meet an interesting person, or drink in the beauty of a natural wonder.

General aviation people know what this means. Distances are less of a problem for them. They can pack up the family for a weekend and travel hundreds of miles, sharing sights and sounds formerly reserved for once-a-year vacations.

No longer are they limited to hearsay accounts. They experience, firsthand, those things that the earthbound only read about. History, geography, sociology and other disciplines are not sterile concepts, but dynamic realities. Add to this a wider vision through which their worlds and ideas grow with each flight.

Education through books? For some, perhaps, this is all that's desired. But it's shallow compared to the lessons learned while being there on the spot.

201
201

HOMECOMING

Thomas Wolfe wrote it in bold letters: "You Can't Go Home Again."

I captured that feeling when, for the first time in over twenty years, I returned to the neighborhood in Pittsburgh where I was raised as a boy.

The houses looked the same, and Little League games were still played on the old baseball field. But some things were different. Most of the neighbors were unfamiliar; my old grade school was now an office building; even the corner drugstore was gone. As I walked down the streets and looked around, I realized I was a foreigner — an outsider who no longer belonged.

So, Wolfe was right, I thought, especially if "home" is a three bedroom house surrounded by grass and trees, punctuated by good neighbors and secure youngsters. For houses can be razed; grass and trees uprooted. Children leave the family; neighbors move or die.

Suddenly, I heard the sound of an airplane flying overhead. When I looked up, it all was brought into proper focus. How foolish of me. How short-sighted.

Looking at the horizontal dimension, my perspective was limited to man-made creations and values. Above was that which transcends — the sky.

Ah, the sky. The eternal sky. Above the steel mills of Pittsburgh, the traffic of Los Angeles, the gaiety of Paris, or the Kremlin in Moscow, it never changes, constant in its laws and promises.

Wherever he may be at the moment, a pilot need only look to the sky as a place where he is no stranger. For others the sky is the limit; for the pilot it's home.

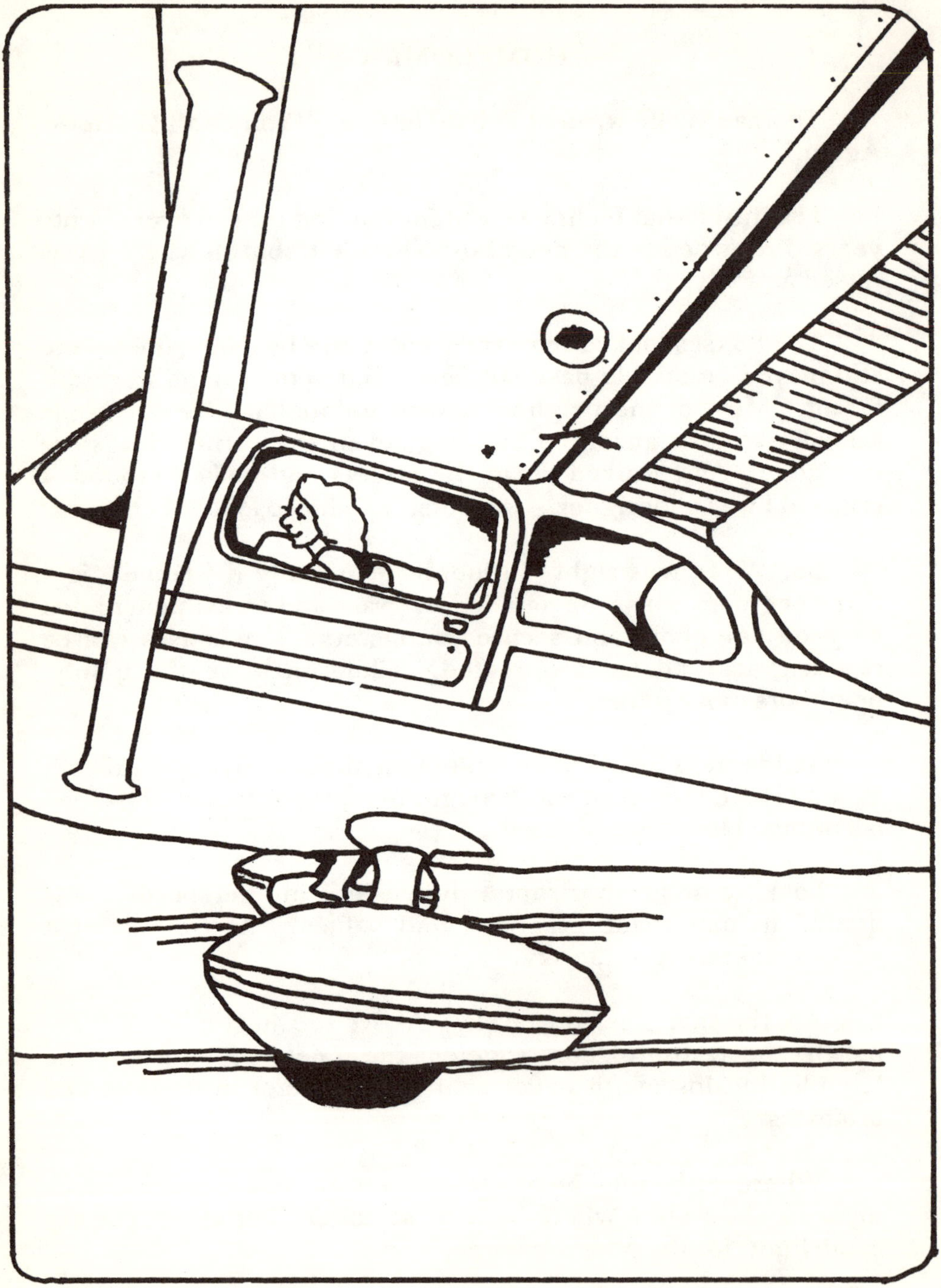

GIFT FOR A LIFETIME

An interesting true story comes out of Dallas, Texas about an oil tycoon who donates thousands of dollars annually to send needy children to summer camp. As a result of his kindness, every year one hundred underprivileged boys and girls are bused from their ghetto sidewalks to the natural freshness of the great outdoors. During a two-week retreat, the young people commune with Mother Nature — a rarity for those who, for the most part, never see as much as a tree throughout the rest of the year.

It's a great experience, all right; unfortunately, it is short lived. After fourteen days, these children are bused back to their depressing environments, and the sights and sounds of summer camp become only memories.

I wonder. Of how much more value are the gifts we give our children? Let's see...we may give them a bicycle..., a stereo..., a new suit of clothes..., even a new car.... Yet, each one, sooner or later, wears out, and something else must take its place.

Last year, a friend of mine bought his daughter a rather unique birthday gift — equal in cost to two weeks of summer camp. He gave her flying lessons — as many as necessary for her to solo.

Sound strange? Indeed, until I realized the potential of his gift.

With flying lessons comes an opportunity for her to grow in a new dimension.

With flying lessons comes a sense of responsibility that keeps her too busy to get into much trouble.

With flying lessons comes a thrill of being one with the mysteries of the sky.

With flying lessons comes a pride as the student accomplishes something on her own.

With flying lessons comes a gift that will remain with her for the rest of her active life.

Flying lessons for a teenager? Why not? They sure beat two weeks at summer camp.

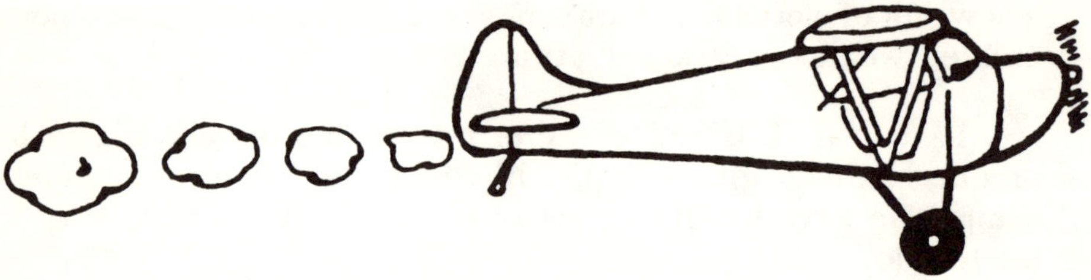

''As long as there exists a fellow pilot...he will never be alone.''

People

← TO ALL GATES
RENTAL CARS →
HAW

THERAPY

I love airports. I enjoy, especially, standing in a mezzanine overlooking the main floor of a terminal that is alive with a bustling crowd.

This is not your usual group of people, much different from those on the streets of New York City or any other major metropolitan area. Airport crowds are not impassionate beings walking briskly, staring straight ahead, alone with preoccupations. These crowds are people full of vigor, wearing expressions of excitement shared only by those about to embark on an adventure.

Look at them. Each has a different story to tell.

There's a teenage girl embracing a soldier who has just returned home from basic training. Next to them another couple leaves for a honeymoon in the Bahamas. There are leaders of business, of industry, even nations flowing as if part of some unseen river. There are children, parents, grandparents — people of nearly every age and occupation weaving among themselves.

But, unlike people in other crowds, these folks acknowledge one another. Some even smile at total strangers. They are aware, it seems, of the common bond that unites them.

One thing is certain. They're not at all impressed with a static existence. These are people on the move — growing in their outlook and appreciation of life.

The next time you feel depressed, or think that zest has gone out of life, visit your local airport, find a good vantage point, and watch the people.

There's a lot of excitement going on out there.

MY PROBLE
ROBLEMS YO
BLEMS OUR RD
MY PROBLE
ROBLEMS YC
YOUR PROB
EMSM S
OUR
P
ROBLEMS M
MY PROBLEM
S YOUR PROBL
LEMS OUR PR
OBLEMS M
YOUR PR
OBLEM

PROBLEMS

Dr. Jim Plinton, amiable representative of Eastern Airlines, tells the story of the time when an 11-year-old lad boarded a jet and began a flight from his home in South America to Toronto, Canada for a year's schooling. His first point of debarkation was John F. Kennedy Airport, New York. As he waited for his next flight, the boy felt a bit dizzy and nauseous, but said nothing.

Upon arrival in Canada, he felt worse — much worse. An alert paramedic suspected something was wrong. The look in the boy's eyes was not quite right. He escorted the youngster to an isolation room and summoned a physician. The doctor, too, was suspicious, and after telephoning the boy's parents in South America, announced that the young man had a highly contagious, rare tropical disease.

As a result, everyone in the plane that landed at JFK and everyone in the plane that left for Toronto had to be contacted and vaccinated.

Hundreds of people who had never heard of the boy's birthplace were now made aware, in a dramatic way, of this remote country and were directly involved with one of the problems of that tiny nation.

At one time governments could afford to be isolated one from another. Our problems were our own. Today, the problems of a remote, sub-tropic nation can become the problems of people throughout the world — chiefly due to air travel.

I wonder if Orville and Wilbur realized what their 12-second flight at Kitty Hawk would mean in terms of this international dimension. Because of what they and their followers have given us, more than ever, we indeed, are our brothers' keepers.

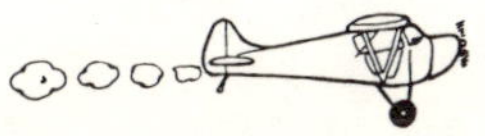

SPEAK UP!

Yesterday afternoon, my neighbor peered over the backyard fence and held high a newspaper clipping about a fatal accident involving a pilot who attempted to fly under some power lines. As a commentary on the whole of general aviation, he snarled at me: "There's too many of you crazy pilots cluttering up the sky. Ain't safe for none of us no more."

I suppose I could have responded with some logical argument or experimented with a display of violence, but I realized that neither would help my cause. So, I dismissed the situation with some worn-out statistical cliche, and returned to planting my petunias.

Now, I wonder. Could this lost soul be right? After all, isn't just one crazy pilot too many?

I wonder, too, if somehow I contributed to the situation. Haven't I seen others bend the rules of safety? I'm not referring to those "tower buzzers" — they're too obvious and, fortunately, few and far between. But I think of the times I see someone polish off a few beers at the airport bar, then invite his buddies for a ride in his Cessna 172. I think of the time I notice another depart an uncontrolled airfield in a haphazard manner without respect to any traffic pattern.

Yet, what do I do? Nothing. Absolutely nothing.

I could approach the offender and politely, but firmly, remind him that he is endangering himself and the rest of us. But I don't. Why? Is it because I'm afraid of his reaction?

The late Martin Luther King, Jr., while commenting on the brutality inflicted on those involved in a Birmingham, Alabama demonstration, lamented: "It was not the cruelty of the *evil* men that disturbed me; it was the silence of the *good* people who stood by and did nothing."

Am I guilty of the same sin of omission?

By speaking up, at least I can contribute something positive to aviation safety. Who knows, I even might be able to get my neighbor off my back.

INDIVIDUALISTS

A psychologist told me recently there are but two kinds of people in the world — conformists and individuals.

Conformists draw their security from society. Success comes from acceptance by others. Failure is a source of embarrassment. Value norms are established by groups.

The individualist is different. His security lies not in others, but in himself. Success is self-determined. A failure doesn't worry him; he knows that uncertainty is an unavoidable factor in life. As far as value judgments, they're his own.

A popular philosophy once espoused the belief: *Vox Populi Vox Domini* — "The voice of the people is the voice of God." But, what if the people are *wrong*? Should we conform to the flow of the masses?

Throughout recorded history are stories of those brave souls who refused to follow the herd and "dared to march to the beat of a different drum." These individuals were not content to be a *part* of history: they *made* history.

Ever since the first flight at Kitty Hawk, a new breed of individualists has been making history. We call them: "pilots." You see, every private pilot — from the Wright brothers to the most recent qualifier for a license — was advised at one time or another by well-meaning friends and family members not to start flying. "It's too dangerous," they said. Screaming newspaper headlines vividly describing accidents coupled with visions of what might happen lent support to their arguments.

Had Orville, Wilbur, and every future pilot listened to the advice of the masses, man still would be standing flat-footed on the soil gazing in awe at the birds in flight.

Unlike conformists, pilots are self-sufficient beings who refuse to be molded by outside forces.

As for me, I'm going to the airport this afternoon and join some fellow individualists and see whose turn it is to make some history.

WORLD WAR I
Fokker Tri-Plane
German
"Red Barron"
Shot
Plane

UPWARD BOUND

For the last five years, something strange has been happening in Florida. High school students, once on the list of probable "dropouts" are now making high marks in college while pursuing noble careers such as medicine, law, and business.

What was the turning point in their motivation to study?

Aviation.

The story is fascinating. Realizing that teenagers are an active bunch, the folks at Embry-Riddle Aeronautical University in Daytona Beach initiated a new wrinkle to the concept known as "Upward Bound."

"Upward Bound" is a nationwide program designed for students who are college material but lack the drive to study and apply their skills in achieving academic excellence. Naturally, some programs are more successful than others, but at Embry-Riddle, the rate of success is absolutely phenomenal. A remarkable 98 percent of the students who complete the "Upward Bound" program are accepted into the colleges of their choice. Think of it — 98 percent potential high school dropouts transformed into college students. And aviation plays a major role.

Embry-Riddle's approach is one of positive reinforcement. If a student attends every class of school for a week, he earns a flight in a Cessna 172. Miss even one class, the student doesn't fly. It's as simple as that.

On top of this, his teachers use aviation terminology to supplement lesson plans. Projects are aviation-oriented. In short, the student becomes immersed in studies and in flying. The result is that he's so busy learning *positive* things that there is little time for *negative* influences.

A juvenile judge once remarked to me: "In my entire career as a judge, I've never had a student of aviation come before my bench."

I wonder. Could it be that one of our best keys to "juvenile decency" is through aviation?

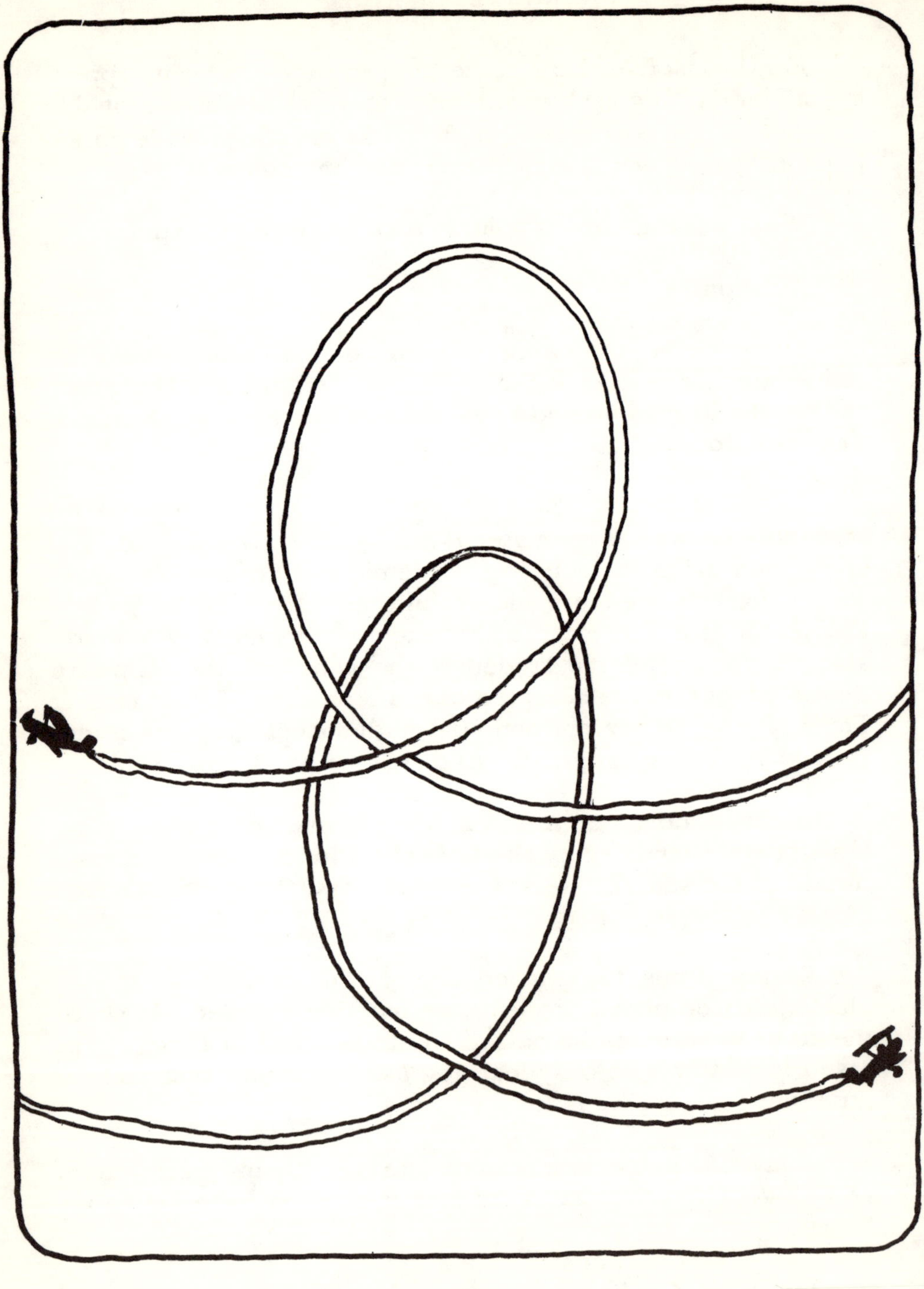

THE FRATERNITY

Gutzon Borglum, the famous sculptor, once made a statue of Abraham Lincoln sitting on a bench.

An admirer said to Borglum: "Why did you put Lincoln at the very edge of the bench?"

The sculptor replied: "Because Abraham Lincoln was just that kind of a man. He did not feel that man can be alone or live alone, and he always left room for someone else."

This is the true nature of man. To be alone, to live alone, to think only of self is to live in a vacuum — and nature abhors a vacuum.

"No man is an island," wrote philosopher John Donne, to which we must add: "and no one should want to be one."

Alone! Man cannot stand to be alone. Most hardened criminals will tell you that there is no punishment equal to that of solitary confinement.

Perhaps one of the intangible blessings enjoyed by a pilot rests in the fact that he is part of a fraternity — a family — of fellow aviators in which each member shares a common bond, and can support the other through a common sense of identity.

In its proper balance, this unique fraternity means that there is always someone to cheer him when he is sad, and rejoice when he is happy. There is always someone who can talk his language and share his problems. There is always someone, whatever his race, color, or creed, with whom he can communicate, and grasp another dimension of this world.

For many, this aviation fraternity is sanity in a sometimes insane world; it is stability in a world of flux and doubt.

No matter who he is, no matter where he is, as long as there exists a fellow pilot, an airport, or even a plane in the sky, he will never be alone.

Then they said, "Don't worry, Mrs. Greene. She'll grow out of it!"

PILOT WITH A MISSION

The early 1940's. It was a different era. "Beatles" were still bugs; when a student went to the principal's office, it was the student who was in trouble; the stock market was going up and women's skirts were going down; the "big war" was exploding in Europe and Japan; and a woman's "calling" in life was well established — get married, have babies, and keep the house neat and clean. At least that was the goal of most women...but not Elizabeth E. Greene.

As a sixteen year old teenage girl flying airplanes, Elizabeth ("Betty" to her friends) broke the stereotype mold.

"It's a passing fancy," said some of her neighbors. "She'll grow out of it."

But Betty Greene knew what she wanted. Uninhibited by traditional roles, she was one of three women students in a class of forty-three at the Government's Civil Pilot Training Program. This tall, slender, Seattle native met every requirement and was selected to fly for the Air Force.

By this time it was obvious that Betty Greene was determined to go above and beyond the call of duty. Flying B-17 bombers while testing military equipment, she became one of the first women to pilot a plane into the stratosphere.

Yet it was after World War II that Betty made her real mark in life. Although other military pilots set aside their log books for careers in business, law, medicine, or one of the "normal" pursuits, Betty reached out to touch other people through aviation.

"If airplanes are used for war," she asked, "why can't they be used for peace?"

At the urging of the Summer Institute of Linguistics, Betty became the first full-time employee of what was to become the United States arm of the Mission Aviation Fellowship (now located in Fullerton, California). Piloting a war surplus Grumman

Duck amphibian biplane, she flew food and medical supplies to people in Peru, and was the first woman to fly over the rugged Andes Mountains.

After a brief stint as an administrator at MAF headquarters, she returned to flying mission runs in Africa. More than once, using "postage stamp" grass strips for runways, she relied on her expert skills in transporting the sick and injured to hospitals sometimes hundreds of miles away, arriving in the nick of time to save a life. "So often we could see God's hand in the timing of things," she said.

Today Betty Greene is still active in directing the activities of the MAF, thus combining her two loves — flying and the work of the church. Although she has had many job offers which far surpass the salary paid by MAF, she explains why she remains in missionary work: "Faith," she says, "is more than just going to church on Sunday."

Betty Greene is not your typical missionary. In fact, she's not a typical person. After all, she is a pilot.

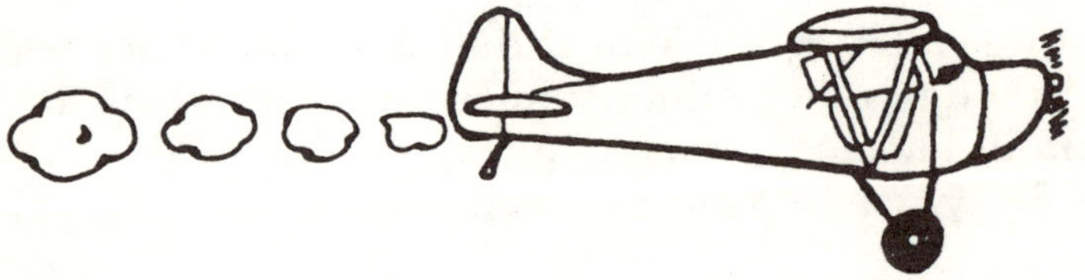

WEALTH

Don Holton is a very rich man.

Recently selling his automobile dealership for a sizable profit, Don Holton is able to retire at an age when most of his contemporaries still struggle up the rungs of the corporate ladder.

But Don's quest is not for the luxuries of an early retirement. Don's passion is flying — a love affair which began when his grandfather took him to an airport to watch Waco biplanes taking hops around the pattern. It's only natural, then, that he should pursue his overt interest in aviation now that he has more time for such things.

But take a spin out to the airport in Ormond Beach, Florida, and you'll likely find Don climbing into the seat of a renovated biplane for a little sightseeing along the ocean beach or preparing for a long trip akin to a fifteen hour venture through rain and snow he took while flying from Maine to Florida a few years ago.

Why would a man who can afford to fly in the comfort of an air conditioned cabin choose to endure the biting wind, whirling dust and smell of exhaust fumes — just a few of the inconveniences of an open-air cockpit?

"Getting into one of those modern contraptions is just 'riding high,' " says Don. "My way of flying."

To those of us who scratch our heads in bewilderment, Don explains that his mother was probably frightened by a Sopwith Camel. This theory notwithstanding, there must be a reason why he, and others like him, continue to brave the elements without the protection and comfort offered through modern technology.

Perhaps it's a way for them to reach back and touch history, or it may be one of a hundred other reasons. Whatever his justification, Don is doing exactly what he wants in spite of the standards set by anyone else.

As I said, Don Holton is a very rich man.

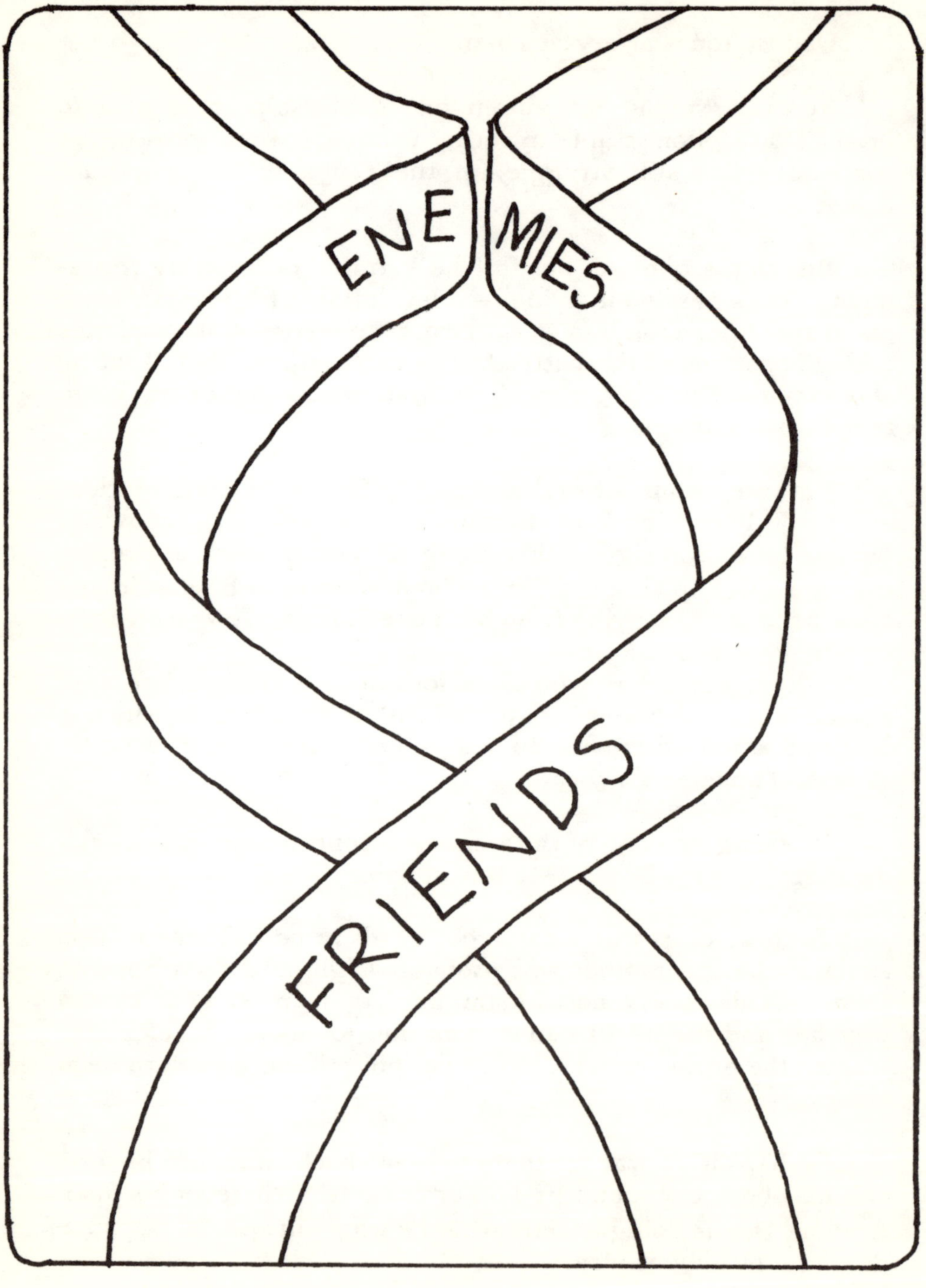
ENE MIES
FRIENDS

THE ENEMY

"Russians."

The name frightened me. Visions of intrigue, world domination, nuclear weapons, and overkill flooded my imagination. Read the headlines. Can the world of Dr. Strangelove be very far ahead?

Consequently, when I was selected to meet with four Russian representatives of Soviet aviation education in the summer of 1979, I was apprehensive, to put it mildly.

Frankly, I didn't quite know what to expect, but I was in for a pleasant surprise. Instead of meeting clones of Ivan the Terrible, I got to know brilliant, sensitive men who were as suspicious of me, at first, as I was of them.

However, it didn't take long for me to realize that they were not ogres; they were people — aviation people — concerned not so much with ideologies as they were about how to increase our odds for survival, whether it be flying left seat in an airplane or living day-by-day in a world resting at the mercy of a red telephone.

Throughout the week, we studied together. We laughed together. We shared common problems. We spoke about safety and precision, awareness and adventure. We discussed more personal things, too, ranging from the difficulties in motivating employees, to the frustrations of a father in raising modern teenagers.

Soon, much too soon, it was over. As the Russian delegates prepared to depart for the Soviet Union, we stood, together, for one last time, alongside their Aeroflot jet and embraced one another warmly.

Their plane took off into the heavens, and I walked back to my car. Then, it hit me.

"What's come over me?" I thought. "I must be out of my mind. I can't carry this love stuff too far. After all, they are the enemy...aren't they?"

FEDERAL AVIATION ADMINISTRATION

AVIATION PEOPLE

I met a friend of aviation recently.

Langhorne Bond, chief administrator of the Federal Aviation Administration, has been cussed and discussed. Recent headlines about deregulation, accidents, and industry disputes haven't made his job any easier. But while sitting behind his desk in his Washington, D.C. office, he must maintain a delicate balance between strict observance of the rules and common sense for the improvement of all dimensions of aviation.

Oh, I've heard the "war stories" about the infamous Mr. Bond. I might have purchased one of the "Impeach Langhorne Bond" bumper stickers had I the opportunity. Yet meeting this person assured me that here was someone who loves aviation and is dedicated to serve its entire community. Breaking through the "Washington-eeze" — a term given to one's ability to use a cluster of words while continually circumventing the real issue — Mr. Bond displays the courage to zero in on the heart of a problem and take action without concern for personal image building.

Once again it becomes apparent to me that aviation is people. Airplanes, regulations, airports, schedules, they're all part of the package. But *people* remain its focal point.

The air aces of the world wars, mechanics, astronauts, ground crews, fixed base operators, barnstormers, corporate pilots, instructors, airport managers, airline captains, and private pilots all share a common pledge to do whatever is necessary to keep the skies safe yet full of adventure.

That's what aviation people are all about.

And Langhorne Bond is one of these people.

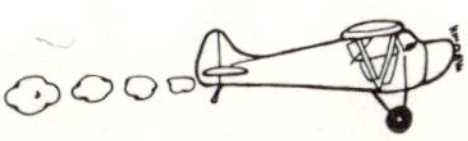

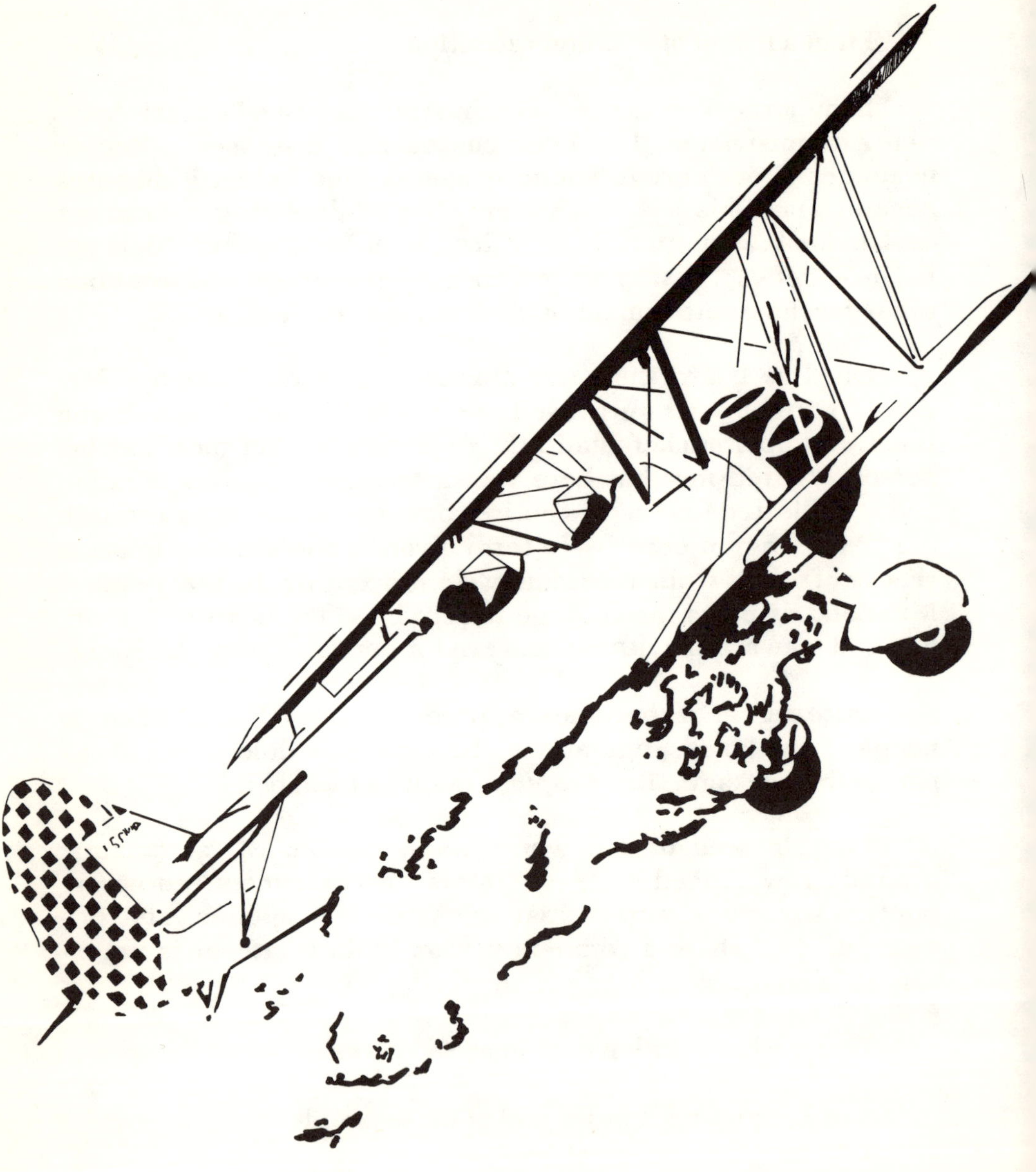

A HERO REBORN

General Paul W. Tibbets is a hero...again. This time, perhaps, without interruption.

Those of us who were weaned on the headlines of World War II remember in vivid detail the one event that rang down the curtain on the fighting — the dropping of the first atomic bomb on Hiroshima.

The pilot-in-command of the famed Superfortress dubbed the "Enola Gay" was the young (by today's standards) twenty-nine-year-old Paul Tibbets. As the "old man" of a crew barely older than most high school seniors, Tibbets not only shouldered the responsibility of leading the air attack, but, for months before, carefully guarded the secret of the bomb.

Paul Tibbets made his historic run on August 6, 1945. At precisely 8:15:17 local time, from 30,700 feet, he ordered the release of the 9,000 pound "gimmick" (as his crew called it).

Never before had the world witnessed such destructive power in one weapon. Although the sun was shining brightly and the bomber was eleven miles from the target, the flash of the explosion nearly blinded Tibbets who was looking away from the blast. Shock waves hit the plane with a force of 2.5 "Gs."

As he headed the plane back to the base on Tinian, Tibbets announced to his crew: "I think this is the end of the war."

Following V-J Day, Tibbets received the accolades of a national hero, including an invitation to the White House for a private audience with President Harry Truman.

But in the 1960s, something happened to change all this. The anti-Vietnam demonstrators, the "Ban the Bomb" advocates and other proponents for peace labeled people like Paul Tibbets as "War Mongers." To them, Tibbets became an unwelcomed reminder of America's "dark days."

Now, thirty-five years later, with the help of books such as

The Enola Gay and *The Tibbets Story*, along with his personal appearances before aviation-oriented audiences, General Paul Tibbets is recognized by most people as one of America's honored heroes.

Today, he is the successful president of Executive Jet Aircraft in Columbus, Ohio. On occasion he reflects upon his military career.

Does he regret his role in the dropping of the first atomic bomb?

''Not for one minute,'' he says.

Then, with a pause, he stares at the ceiling, and adds a touch of irony: ''Yet, we must find a way to eliminate war as a means of settling quarrels between nations.''

"Airplane and I were one."

Precision

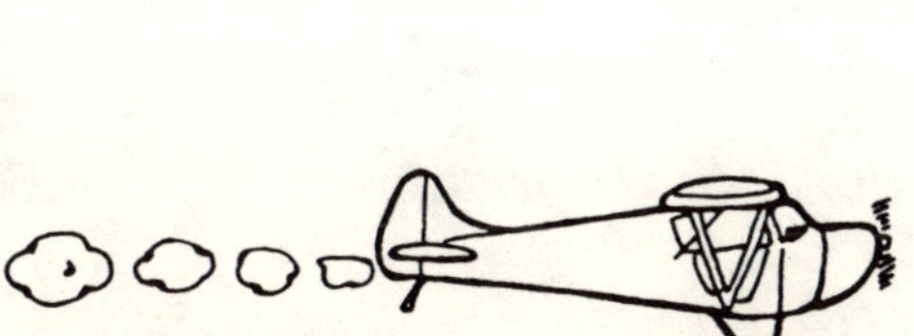

MASKS

The playwright, William Shakespeare, coined the phrase: "All the world's a stage, and all the men and women merely players." Unfortunately, for the lot of humanity (and this includes pilots), the average being often is unsure of which role to play, and he chooses, instead, to hide behind a variety of masks: one for his boss, another for his parents, or his professor, or his clergyman, or the F.A.A. Examiner, or his best friend, or a potential customer. All masks — each one different.

In fact, we all have a collection of masks, but the private pilot has a unique assortment. Consider, for a moment, the four most prevalent:

1. The mask of *Self-Deception* when he insists: "Waddaya- mean 'check ride'? All airplanes fly the same."

2. The mask of *Overconfidence* or the "You-Can't-Tell-Me-Anything-I-Don't-Know" syndrome.

3. The mask of *The Super Pilot* that leads him to believe that it is the reckless abandon of the legendary barnstormer that makes him a *real* pilot.

4. The mask of *Wrong Priorities* whenever he feels he must reach home on time in spite of unfavorable weather reports.

Masks — the pilot can ill afford to fly while wearing them, because the only one he fools is himself.

On your next preflight inspection, take just a few minutes to remove all masks before climbing into the cockpit.

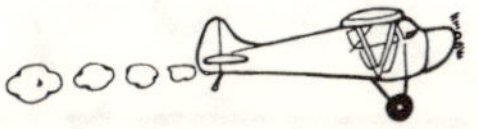

LAW
LIBERTY

THE BALANCE

Have you ever observed the reaction of one of your earthbound neighbors as he watches a private airplane overhead appearing to dance among the clouds? Chances are he'll watch for a few moments in silent reverence, then exclaim something such as: "Man! You guys are lucky to fly like that — free as a bird."

Free as a bird? I suppose we might agree with him. But that's only half the story, isn't it?

In a sense, a private pilot has absolute freedom. After all, can't he go wherever he wants without restrictions or traffic jams or speed limits? Isn't he free to transform day-long journeys into side trips of a few hours? Doesn't he enjoy the power that comes with this freedom to put his airplane into configurations which boggle the mind (or a passenger's stomach)?

Freedom? You bet.

At the same time, every pilot knows he is bound by absolute obedience. He must conform to the laws of nature — laws set in concrete during the days of creation. If the pilot doesn't believe this, watch what happens when he disobeys these laws.

Absolute freedom and absolute obedience — both create a healthy balance for any pilot.

Freedom gives him god-like power; obedience keeps him human.

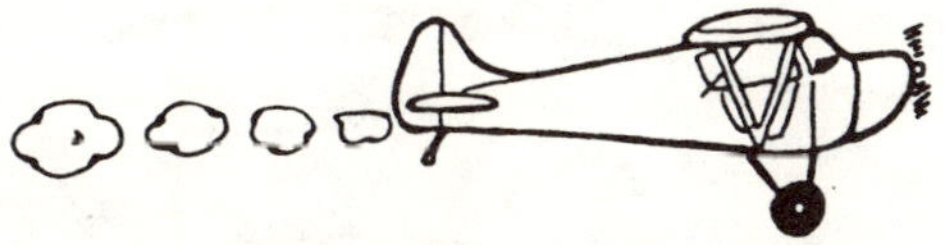

R.I.P.
SHOW-OFF

THE MACHO PILOT

Ah, to be a ''macho man.'' Wouldn't it be wonderful? Think about it. Men would follow you, women would dream of you, and young children would worship you.

Exaggerated? Let's look at the facts.

As a pilot, you have a built-in ''macho'' appeal. People look up to you as a defier of the odds, a daredevil bent on adventure. Visions of the white-scarfed, goggled fighter pilot dance in their heads. You brave the elements, and conquer them.

The ''macho pilot.'' What a picture.

Oh, sure, the image may be blown out of proportion by someone unfamiliar to general aviation. Those of us who fly know that the one who sits in the left seat of a Cessna 172 can be a typical office worker, housewife, sales clerk, student, or teacher. He doesn't look different from the norm. His basic personality doesn't change with a pilot's license.

At the same time, society (whatever that term implies) places a subtle demand upon the private pilot to ''show the stuff he's made of.''

For example, imagine that you and your best girl are ready to leave on a cross-country Sunday afternoon flight. Undoubtedly, she's impressed with your ability to navigate an airplane through the skies. No doubt, too, some of her admiration for you is directly related to your flying skills. But, weather advisory shows severe thunderstorms en route. What do you do?

Common sense tells you to wait for the weather to clear, or to postpone your excursion to another day. But, no. That would be admitting weakness, wouldn't it? So, urged on by the demands of society (that nebulous word, again) and the urge to prove to your girlfriend that no ''little ol' rain'' is going to stop you, you forsake the warnings, push the throttle to the wall, and head straight for your destination. After all, you're a pilot. You wouldn't do anything to tarnish that image.

That's "macho."

The next day, your family and friends are shocked by the news of your untimely death from the accident.

The N.T.S.B. report is not yet in, but everyone knows the reason for the crash.

At your funeral, no one mentions this. Instead, each pays tribute to your memory.

"He was a nice guy," says one.

"Adventuresome," says another.

But, it was worth it. For after you are placed into your grave, someone will inscribe on your tombstone: "Here lies a 'macho pilot'."

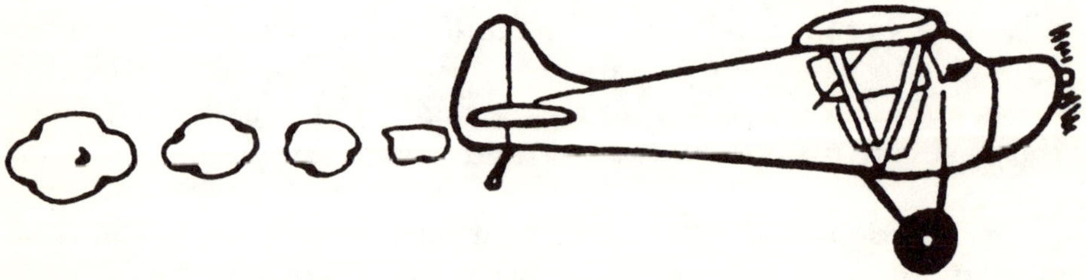

ONE THING

Just ''one thing'' can make all the difference in the world. It did for New York Yankees' catcher Thurman Lee Munson.

While attempting to land his private jet aircraft at a local airport, Munson and his two companions fell short of the runway causing the plane to burst into flames. The two passengers were able to escape to safety without serious injury. Thurman Munson, however, was killed.

What was the difference between Munson and the other two on board? Investigation by those at the crash site revealed that the two who escaped wore both the lap belts and shoulder harnesses; Munson wore only his lap belt.

One thing, that's all.

It takes no more than one thing to alter human events beyond the expected. *One* bent tooth of the tiniest gear can render an expensive watch useless. *One* nail lying innocently on a highway can cause a tire to blow, hurtling a car from the road. *One* grade point can make the difference between passing and failing.

Air safety has often resulted from *one* thing — one visual check of the oil...one call to a flight service station...one close inspection of the control surfaces...one test for water in the gasoline.

The difference between triumph and tragedy, between life and death, may boil down to just *one* thing.

For Thurman Munson, that one thing was a shoulder harness.

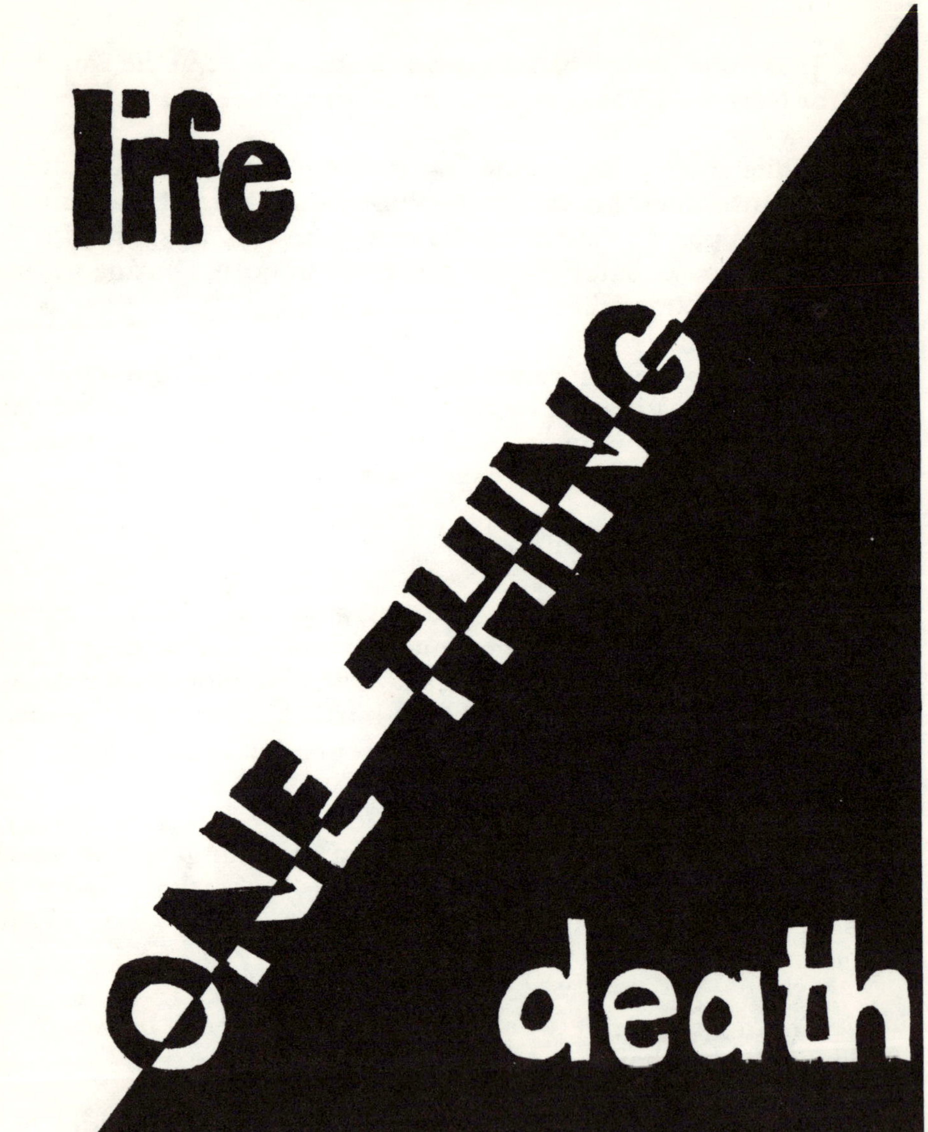
life
ONE THING
death

A MARRIAGE

It was love at first sight.

Physically, she was nothing much to look at. She wore telltale signs of age, clearly showing that she fought her share of tough battles. But, to me, she was absolutely beautiful.

Yet, I guess everyone feels the same about the first airplane he owns.

Did I say "owns?" Forgive me. No one really "owns" an airplane. He may be entrusted to care for one or to nurse one back to health, but never to "own" one like he would a piece of furniture.

Perhaps this is why a new buyer is apprehensive, and it's not because of the cash and contracts and finance charges; these are the mundane business matters through which he must work at the present. Rather the new buyer is nervous because of the tremendous responsibility that comes with this commitment to his airplane.

Now, I shared that feeling of uncertainty.

Over the next few months, Airplane and I flew the skies together, getting to know one another. Yet, in keeping with the spirit of her ancestors, without warning she would see just how far she could stray before I caught her. A slight deviation in direction...a jolt here and there...a change in altitude....

"Oh no you don't," I'd have to say to her. "Get back to where you belong." Only then, somewhat reluctantly I might add, would she slide back on course.

On one gorgeous Sunday afternoon, Airplane and I were flying at 3,000 feet over Michigan above the shores of Lake Huron.

At that altitude, the wind was blowing over the water at 40 knots. Heading directly into the wind, I pulled the throttle to idle, and lowered the flaps until we hovered, suspended in mid-air, without moving in any direction. Together, Airplane and I were in total harmony with the environment.

The sensation was magnificent, and I was thrilled.

Airplane liked it too.

I believed more than ever before something told me by an astute observer: ''Just because airplanes may be inanimate does not necessarily mean they don't have souls.''

After several minutes, I applied the throttle, dumped the flaps, turned 180 degrees, and headed home. But from that moment on, Airplane and I were one. It was a fantastic marriage.

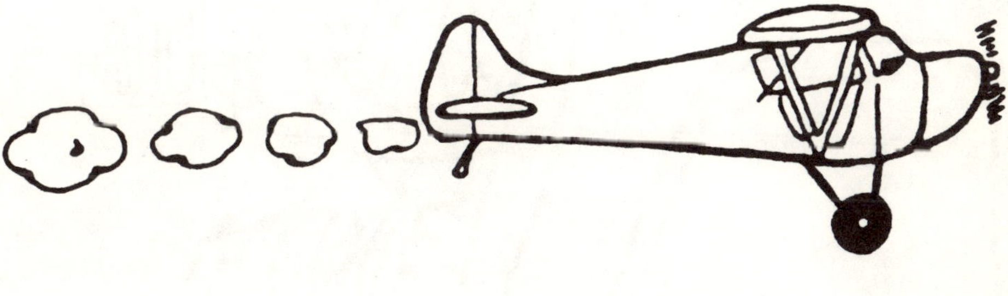

PRO

PROFESSIONALISM

Ever ride jump seat in a Boeing 727?

For this private pilot, it was a thrill — a peek into the big leagues of aviation.

The captain, co-pilot, and flight engineer responded as a team as if directed by some unseen conductor. Taxi and takeoff were "by the numbers." The co-pilot called out the airspeeds: "V one...V two...Gear up...." Everything went perfectly.

Within minutes, when we reached cruise altitude, the mood inside the cockpit changed. The conversation became relaxed. "How's the wife and kids?" "About that Tripacer I want to buy...."

Don't get me wrong. Each man was alert to his responsibility. The crew members constantly monitored the instrument panel and maintained communication with approach control. Eyes continually scanned the horizon for other aircraft. Yet they stayed "loose" and enjoyed every minute of the flight.

As we neared our destination, once again it was all business. The co-pilot read aloud the checklist, not missing one point. The landing was smooth — very smooth, indeed. I think I detected a look of pride in the captain's smile.

On the surface, it might appear as though there was a sharp contrast in attitudes among the crew members. No so. Each member of the team maintained a balance of precision and relaxation. Precision in that each step was executed as prescribed. Relaxation in the event that if something went wrong, the men could respond calmly and deliberately without panic.

To keep this balance — that's professionalism.

FLYING
POP
The Creep Your Daughter is Dating
Asking Boss for a Raise
POP
New Dent in Car
POP
Vacation Plans
POP

NARROW-MINDED PILOTS

Since childhood we have been told about the virtues of remaining open to new ideas, to be tolerant, and to be receptive to the opinions of others. It is admirable, so we've been told, to be broad-minded. However, may I suggest that there is an immeasurable value in the posture of being *narrow* minded.

For example, when I take my car to a garage for repair, I don't want a mechanic who replaces everything in sight until, by mere chance, he happens to discover the problem. I want a *narrow*-minded mechanic who knows exactly what's causing the problem, fixes it, then sends me on my way.

When I experience a malfunction about myself, and a physician's diagnosis spells "surgery," I want a *narrow*-minded surgeon who knows exactly what he is going after, so he can cut it out, and leave the rest of me intact.

Likewise, there is much to be said about the *narrow*-minded pilot. In fact, I want a *narrow*-minded pilot who, when flying, thinks only about flying, not about the problems at the office or the latest argument at home.

I want a *narrow*-minded pilot who, in any emergency, knows precisely what to do, and executes it with confidence.

I want a *narrow*-minded pilot who is critical of himself, striving to maintain the exact altitude along the exact course as planned.

I want a *narrow*-minded pilot who keeps in focus the precision demanded of everyone who voluntarily invades the sky.

Broad-minded? That may be all right for the earthbound. But for those who love flying, the art of being *narrow*-minded may be a real lifesaver.

FLYING RABBITS' FEET

So things have been going rough lately. Your recent landings have been less than perfect. Your airplane's annual inspection reveals a run-out engine. Lately, even the weather is not cooperating. You want to cry out for help. Don't panic. Help is as close as your nearest rabbit's foot.

Superstition, my friend. The Word is: "S-U-P-E-R-S-T-I-T-I-O-N."

"Mere folklore," cries the scoffer.

"Obsession neurosis," says the psychiatrist who just walked under a ladder and had to take a "downer."

But those who are privileged to know the power behind these unexplained phenomena do not treat the matter so lightly. It has taken nearly seventy-five years and quite a few flights around the pattern to build up credence in these superstitions.

What particular superstition will help you? That's impossible to predict. What works for one pilot may not work for another. A snake oil salesman would have a field day at the local airport. However, there are some superstitions that most pilots find helpful.

Did you know, for example, that it is bad luck to have your photo taken before a flight, and that you never, never fly on your birthday? Have you ever seen an aisle number 13 on a commercial airliner?

On the other side of the lucky silver dollar, seasoned pilots attest that good fortune follows those who shake the wings of an airplane as the last step of the preflight, and claim that a safe flight is guaranteed as long as you remember to take the same flight bag and/or clipboard on every flight.

There must be more — plenty more. Perhaps you know of some not mentioned here. If so, drop us a note. Keep us informed. The world must know the truth about such things!

Rattle
Crunch
Crinkle

AIRPLANES AND ERASERS

One luxury that comes with writing this book is the chance to correct a mistake in judgment. For example, if I happen to choose a wrong word, I can reach for my well-worn eraser, rub it out, and write another in its place. Before the final draft, I may repeat this process several times. But that's no real problem, since I have a second chance...a third chance...a fourth chance. In fact, I can continue writing and erasing until I get exactly what I'm looking for.

When I used to play baseball and attempted to field a ground ball, on occasions — too many to number — I made an error. Yet, I could erase that memory from my mind, and concentrate on the positive thoughts that there was another inning... another game...another season.

If I fail in school and must repeat a course, or in business and lose most of my fortune, or in marriage and lose part of my family, I find comfort in realizing that there are other schools... other businesses...other potential mates. In each instance, I can erase my bitterness, rise from the ash heap, dust myself off, and begin anew.

Ah, but flying is different — vastly different. How many second chances do I have if I forget to remove the gust locks from the empennage? Or fail to heed the "never exceed" limit on the airspeed indicator? Or jump into the cockpit immediately after downing a few beers at the neighborhood pub? In these instances, I have only *one chance to fail*!

Think of it. Only one chance. I dare not make a mistake.

But, that's the difference between writing and flying. Airplanes contain no erasers.

PILOT
IN
PROGRESS

STUDENT PILOTS

Once a pupil asked a rhetorical question of the Danish theologian, Soren Kierkegaard: "You are a Christian, aren't you?" To the surprise of the youngster, Dr. Kierkegaard answered: "No." Then, he added: "I am in the process of *becoming* a Christian." Kierkegaard explained that religious growth does not cease with conversion, baptism, or church membership. "Becoming" a Christian is an everyday experience.

Perhaps, with a tinge of pride, we inform those we meet: "I'm a private pilot." The more sophisticated may list graduate credentials — instrument, commercial, or multi-engine ratings.

But, to parallel Kierkegaard's insight, every pilot holds only one license — a student certificate.

Woe be unto him who thinks he knows all there is to know about the unlimited sky. Not only is he a potential danger in his complacency, but he is missing an awful lot of adventure.

The serious pilot rests not on his laurels, but looks ahead to other horizons.

He anticipates problems by asking himself: "What if...."

He reviews emergency procedures regularly.

He remains alert to possible malfunctions.

He dares to look beyond the obvious.

The pilot is always a student in his quest for higher ratings, in his zeal for understanding, in his awareness of safety, and in his willingness to grow in every dimension of aviation.

The reason is simple. He is in the process of "becoming." And that makes him a student — *cum laude*.

WHO DOES YOUR THINKING?

The pilot was quite proud of himself. It was his first attempt at landing a 707. He was cleared to runway 27L, and, so far, everything was perfect — the rate of descent and airspeed were "by the book."

Suddenly, at about the inner marker, without warning, both right engines quit, simultaneously.

The pilot grabbed for the microphone shouting: "Tower! I've lost the right engines...both of them!"

"Prepare for a 'go-around'," instructed the controller.

The pilot pushed the left throttles to the wall. But, instead of climbing, the airplane swerved sharply to the right, heading straight for the control tower. The pilot froze at the wheel.

The world seemed to collapse with the ear-shattering crash! Then silence — a dead silence.

"Wh...what happened?" asked the pilot now drenched with perspiration.

"You've just killed over one-hundred people," was the caustic reply of the controller.

The exhausted pilot unfastened his seat belt, and climbed out of the simulator, thankful that nothing was hurt but his ego.

"Today it was a simulation," reminded his wise instructor; "next time it's for real."

Yet, our novice pilot knew that he didn't have enough experience to make this airplane climb with only two left engines operating. He knew, instead, that he should have continued his landing approach. But, because the controller had said: "Go around," he did...without thinking.

The pilot realized that his biggest mistake lay not in his lack of skills with the aircraft, rather in the fact that he failed to stand by his convictions. He let someone else do his thinking for him.

Whether inside or outside the cockpit, we are required to search for the facts and act in harmony with them. In all cases, we must commit ourselves to what we know is best in spite of the advice by others.

In these important moments, who does your thinking for you?

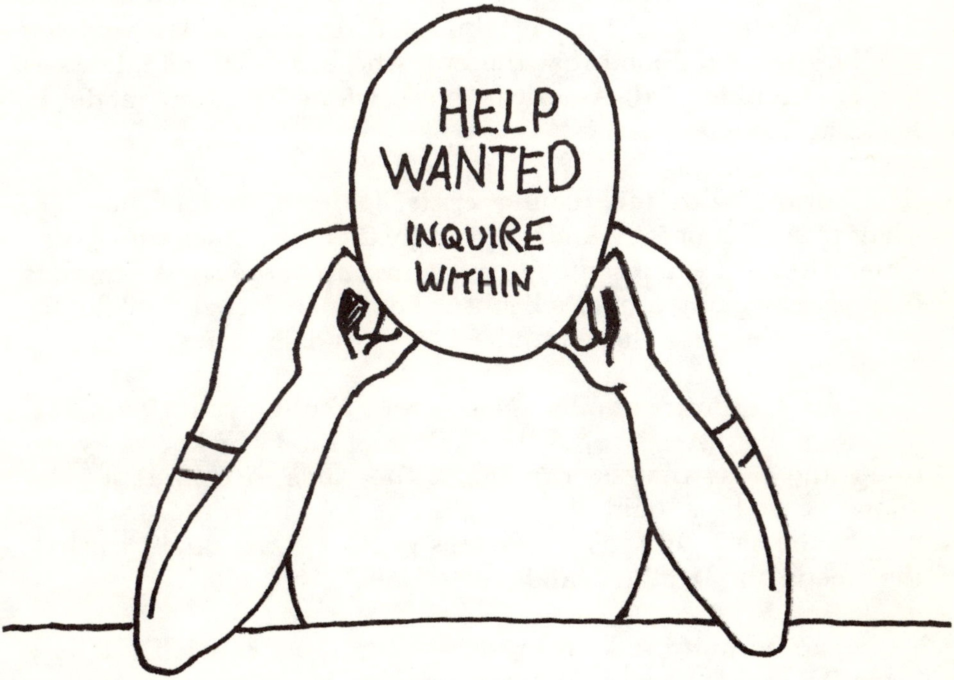

HUMILITY

"The trouble with you pilots," intoned a semi-intoxicated party-goer at a recent neighborhood bash, "is that you're all too conceited. I've never met a humble pilot."

Aside from the fact that he was one of the reasons why I avoid cocktail parties whenever possible, his candid observation had a ring of truth, although, quite probably, he was unaware of his insight.

He was right, of course. Pilots are not humble — in the accepted sense of the word, that is.

"Humility" to most of us equals shyness. We chafe at those who say or imply: "I'm certain I can do this." We've been taught since childhood that the one who boasts of his talents is not the humble, "all-American boy" who's been our model in books and movies.

I agree with my reeling critic. Pilots are not humble. They're confident they, indeed, can fly their airplanes with precision. After all, when someone measures up to the rigid demands of anyone seeking a pilot's license, he displays a set of skills far more complex than those required by most disciplines.

I want a confident pilot. Not "cocky," mind you. Confident. There's a big difference. A "cocky" pilot thinks he knows everything and feels that he can break the rules; a confident pilot knows what he should and should not do within the laws that govern the sky. It's his awareness of his "cans" and "can'ts" that keeps him humble...and alive!

True humility is knowing the truth about himself. It is a virtue by which he does not esteem himself to be more or less than he really is.

I tried explaining this concept to my inebriated partner in dialogue. Alas, my efforts were in vain, and he looked at me with the disgust of an untipped waiter.

''Bah,'' he sneered. ''You're just like all the rest.''

Perhaps he's right. But be he drunk or sober, I'm afraid he'll never understand why.

I
II
III
IV
V
VI
VII
VIII
IX
X

TEN COMMANDMENTS FOR PILOTS

Let us suppose you had the opportunity to counsel a new pilot who has just earned his wings. What advice would you give? What specific "dos" and "don'ts" would you stress?

In harmony with a tradition begun thousands of years ago, permit me to suggest the following "Ten Commandments for Pilots":

1. Thou shalt not mix alcohol and flying time.

2. Remember to file thy flight plan and to preflight thy aircraft before ever turning on the ignition.

3. Honor thy instructors that thy days may be long in the sky which the Lord hath created for thee.

4. Respect thy airplane — its potential as well as its limitations.

5. Thou shalt not fly into thunderstorms, ever.

6. Thou shalt not enter thy cockpit whenever angry, preoccupied, or tired.

7. Thou shalt communicate at all times with those below trying to guide you safely on your way.

8. Thou shalt not hesitate to declare an emergency when appropriate.

9. Thou shalt constantly upgrade thy flying skills through advanced study, higher ratings, and overt awareness.

10. Above all, thou shalt enjoy thyself to the fullest and count thy blessings while flying.

MY PAL, THE FLIGHT INSTRUCTOR

Listen, all you flight instructors who want to be my ''pal!'' I wish for one moment that you would stop this ''buddy-buddy'' routine, scrap the ''Mr. Nice Guy'' image, and do the job you're being paid to do.

Your job is to make me sweat. Becoming a good pilot is a demanding, difficult enterprise. It is painful to have my ignorance exposed. It is mentally and physically exhausting to read charts, copy clearances, maintain course and altitude, and comply with instructions all at the same time. But I better be able to do this — and more — when I'm the victim of an unexpected frontal system.

Your job is to criticize. Not all of my landings are good; in fact, some are downright terrible. If that's the case, tell me so! Such criticism, when well-founded and constructive, is in no way degrading. In reality, it is only through criticism that I can learn.

Your job is to evaluate. Grade my performance accurately, and, if necessary, fail me. If not, my flight training deteriorates to the level of the ''Caucus Race'' in *Alice's Adventures in Wonderland* in which everyone starts when he wants, everyone finishes when he wants, and everyone is a winner. Can there be any sense of accomplishment in this kind of victory?

Look, I can find plenty of ''pals'' anywhere, but where can I find a good flight instructor?

I don't care if you make me love you, hate you, envy you, admire you, or curse you, but please let me *respect* you.

You can be a coach, a father image, a priest, a counselor, a tyrant, a dictator, or all of these rolled into one, but, when instructing me in the art of flying, please, please, please, don't be my ''pal!''

Nobody gets hurt, see?

THE VICTIMS

Yesterday, a crime was committed, and nobody did a thing about it. In fact, no one really seemed to care.

The criminals were not your run-of-the-mill hoods, but were intelligent, well-bred, and clean-cut. The ring leader was a trained professional.

The caper began when a private pilot, with about 250 hours total flying time, drove his car to the local FBO for his 10:00 a.m. appointment — his biennial flight review.

As the pilot neared the airfield, he grew more and more apprehensive. For nearly two years, now, he had rented planes from this FBO. He knew the instructor quite well — even shared a few beers with him — but had not flown with him for 18 months.

He rounded the last bend and parked his car in front of the office. He sat there for a minute thinking to himself: "What if I 'foul up?' Will I be grounded? Am I as sharp in my navigation? Can I still make that turn around a point? What about those emergency procedures? And those new regulations?"

It was 10:00, exactly, and he went inside.

"Excuse me," he asked the girl behind the counter. "Is Joe here?"

"Yeah," she said, "he'll be right with you."

Our friend took his seat and waited. More anxiety.

"All set?" Joe's voice came from the doorway leading into the hangar.

"Sure," he answered, sounding as confident as he could.

"Let's say we go for a touch and go or two. That OK with you?"

"Why...uh...sure." The pilot smiled with relief.

And that was it. Two trips around the pattern. The instructor signed the log book, took $15, and apologized to the pilot that he had to put him through this "necessary evil" as required by the F.A.A. The others standing in the office nodded in agreement.

But, wherein lay the real "evil"? With the instructor who took money under false pretenses? With the pilot for accepting such shoddy treatment? With the F.A.A. for allowing a system such as this to exist? Or was it with those who watched all this and said nothing?

If this was a crime, there's got to be a victim. Was there? You decide. Could it be the pilot who was cheated of an opportunity to have an honest appraisal of his flying skills? What about the insurance company that invested a risk in the airplane? Would it possibly be a future fatality filed neatly away in some bureaucratic storage cabinet under the heading: "Pilot Error"?

Right now, I'm especially concerned. Next month I'm due for my biennial review. I wonder if, at that time, I'll be tempted to allow another crime to be committed.

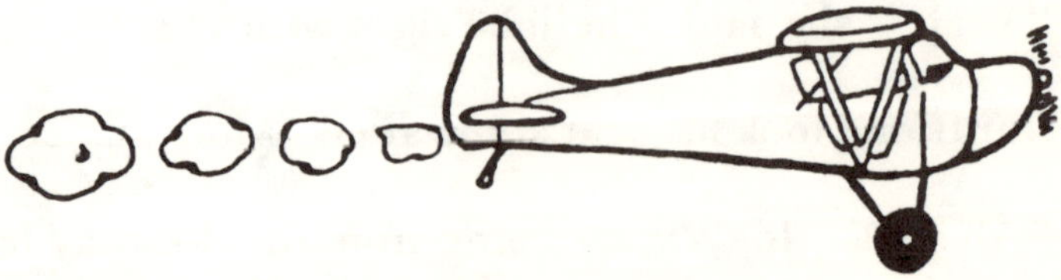

"My son, learn to fly."

Dreams

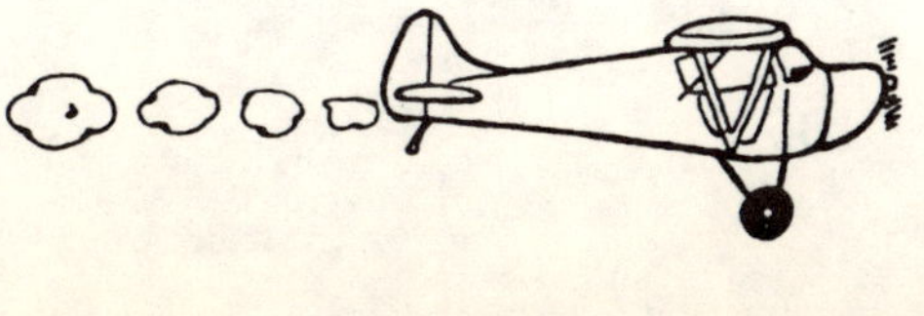

A LETTER TO MICHAEL

My son:

What shall I wish for you — money, power, secure job, happy home? All these may be yours, yet you could miss a joy known by a precious few.

My son, learn to fly, and share the thrills of those who dare to rise above the everyday horizons. This new, higher dimension will cause you to search beyond your own little world to bigger, bolder adventures and insights.

My son, learn to fly, and capture the pride of self-discipline. Learn from your instructors. Practice — daily, if possible — the mental and physical skills needed to glide and soar in harmony with the basic laws of the sky. Continue to think clearly even in the most perilous times. Listen to the voices of experience, for there is much — so very much — to learn.

My son, learn to fly, and join the fraternity of those whose history spans the centuries that began with the first wistful glance at a majestic bird in flight, through Kitty Hawk, to the penetration of man into interstellar space.

My son, learn to fly, and for this encouragement, I ask no thanks — merely a pledge that you regard flying as a sacred trust. Don't misuse it or treat it lightly. Handle it with reverence and awe. If so, someday, with unequaled joy, you, also, will have the privilege to say: "My son, learn to fly."

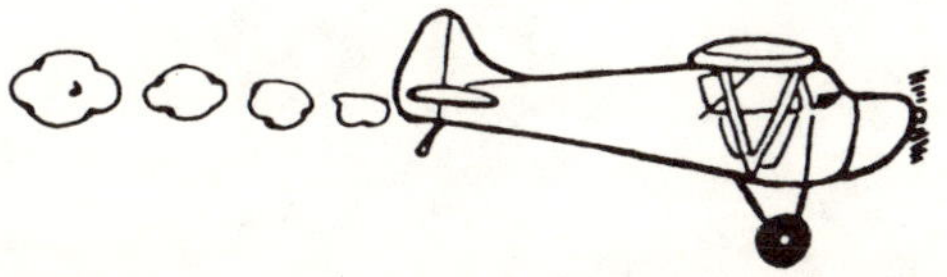

SECURITY
DREAMS

VALUES

Last week I was jogging with a friend of mine along the ocean in Daytona Beach. He was telling me, as he had at least a dozen times over the past year, how he envied those who earned a living through a career in aviation. "Now that's something (puff) I could sink my teeth into," he said.

In spite of his obvious enthusiasm, he never got beyond the wishing stage.

"If I had to start all over again (puff)," he continued, "I would operate (puff) a small fixed base operation. That would be fun."

"What's stopping you (puff) from doing it?" I asked.

"I (puff) already have a good job (puff). I guess (puff) that's it."

I nodded. He had a comfortable position at a local bank. Assistant vice president, I think. And his job was demanding, leaving him little time to enjoy flying or just to hang around the airport.

"If (puff) I stick with my job long enough," he added, "I might (puff) be able to buy a used Cessna (puff) for my very own. But," he said with a chuckle, "I'll probably have (puff) no time to fly it."

We laughed, and our discussion ended there. He would be crazy to think about leaving the bank to gamble in the cut-throat world of private business — especially in something as unpredictable as aviation. Why should he exchange an eight-hour-a-day bank job for a twelve to fourteen hour workday, seven days a week, in some isolated hangar? Where's the security in that?

My friend still flies on occasion — sometimes as much as three hours a month — and enjoys every minute of it. Yet, he'll never make a living in aviation, even though it's his dream.

But, after all, isn't security much more important than dreams?

BABEL

THE INTERNATIONAL LANGUAGE

"Peace" is an "in-word" these days. A tumultuous nation seeks tranquility and forces a Shah into exile. A President attempts to settle a four thousand year dispute. Wise men still follow the star that promises world peace.

Only one thing is missing — a universal vehicle of communication between nations and cultures.

I suggest that those of us who share the freedom of aviation have access to an "international language" which can serve as a basis for better understanding.

It matters not if his name is Jones, Himmler, or Roskovitsky, from the time a pilot turns on the ignition of his airplane, the same rules apply. If you don't believe this, discover what happens when you let the airspeed fall below the stall warning. Watch the wings, if you dare, as you exceed the stress limitations during a power dive.

Defying the Tower of Babel, pilots communicate in a global language unique to their fellowship. Language? Perhaps it would be more correct to say *languages*. For there is the language of awareness...the language of adventure...the language of freedom...and the language of absolute obedience. Any pilot can understand them all, and no interpreter is needed, just as any pilot can appreciate the thrills of flying his first solo, breaking the sound barrier, or seeing his own child begin flying lessons.

In fact, within this sophisticated world of advanced technology, aviation is one of the few remaining universal languages.

Let's face it. Peace is too important to place in the hands of politicians. It is not too far-fetched to imagine that if we let the pilots of every nation talk with one another in the international language of flight, our problems would vanish into the thin air that we share.

But, we're just kidding...aren't we?

BUMBLE BEES
Attention!!
IMPOSSIBLE!
WHAT?
OH NO!
BONK!

THE IMPOSSIBLE

We don't know who he is; unfortunately his name has been left out of the history books, but in 1875, the Director of the American Patent Office submitted his resignation to the Secretary of the Board of Trade because "we have nothing more to invent." According to this expert, it was impossible to produce anything new.

As a result of his fighting in fruitless battles with the Southerners, a German named Zeppelin urged industrialists to manufacture a dirigible balloon. The French Academy of Science scoffed at this as being akin to the squaring of the circle — impossible.

The idea of a heavier-than-air machine propelled by rockets was presented by Herman Gaswindt to the German War Minister. After reviewing the fifth blueprint, the War Minister suggested that Gaswindt be "bumped off" for promoting such an impossible notion.

A firing squad eliminated a Russian, Kibaltchich, who believed in the same impossible theory.

Not too long ago, American scientists declared that it is aerodynamically impossible for a bumble bee to fly. Unfortunately, no one told the bumble bee.

And so they continue *ad infinitum* — what we know as some basic elements of flight never could have been realized simply because others said they were "impossible."

Last week, one of my students claimed that, one day, man will be able to fly even without benefit of an airplane. Now, that's impossible...isn't it?

Floating Metal? Ha!

IF ARISTOTLE COULD FLY

Sometime during the fourth century B.C., more than 2,000 years before Orville and Wilbur Wright conquered the force of gravity, a renowned Greek teacher named Aristotle penned the famous model for modern logic:

All men are mortal.

Socrates is a man.

Therefore, Socrates is a mortal.

Pure and simple — since no one could deny the first premise, the rest had to follow.

Now, let's apply the same line of thinking to aviation:

All metal will not float in air.

Airplanes are made of metal.

Therefore, airplanes will not float in air.

That makes sense, too. For centuries, this Aristotilian logic kept man bound to the earth gazing in awe and wonder at the soaring eagles, and asking himself: "I wonder...?"

Fortunately for Orville, Wilbur, and the rest of us who follow, someone dared to challenge the first premise, and proved that, if designed properly, with enough air flowing over dihedrals, such things as metal, wood, and plastic *could* float through the air.

In fact, if Aristotle, himself, were able to fly his own airplane, he would witness, first-hand, the fallacy of the first premise. He might even discover the secret that the love of flight goes beyond the obvious limitations, and, through technology, fashions its own logic on a higher dimension.

Now, about the premise that "all men are mortal...."

N

PILOT'S CHRISTMAS LIST

Dear Santa:

This year I have a large order for you to fill. My Christmas list may be a bit unusual; nonetheless, I give it without reservation.

Give to all *Pilots* a thirst for awareness of the rigid demands of flight. At the same time, give them an appreciation for their view of the world unknown by their earthbound neighbors.

Give to *Air Traffic Controllers* a skill to communicate effectively with those whose safety depends upon their guidance. Give them a firmness when issuing commands and a calmness in the event of an emergency.

Give to *Airport Managers* a sense of alertness to the needs of the industry as well as the concerns of the community. Give them an open ear to honest complaints and an open mind to innovative ideas.

Give to *Mechanics* a pride in their work. Give them a sense of obligation to insure the safety of those who use their airplanes. Let them never be satisfied with second best.

Give to *Students* a desire for knowledge in every dimension of flight. Give them determination to continue their instruction for as long as they hold a license to fly.

Give to *Instructors* the patience to guide and mold their students in a way that encourages them to become conscientious pilots. Give them fairness to criticize bad habits and to praise striving for perfection.

Give to *Retired Aviators* everywhere pleasant memories they can relive each day. Give them happiness without the fear and cruelty of age.

Finally, give to *All of Us* a commitment to do whatever we can to promote the cause of aviation through our words, through our work, and through our lives.

DEED
DIPLOMA
TITLE
Birth Certific
Marriage
Licence

DIMENSIONS

Why would anyone want to fly an airplane? Not "because it's there" — the reason given by mountain climbers for ascending to the tops of the world. The airplane is not "there" until put there by a pilot.

Why would anyone deliberately place himself at the mercy of unseen wind currents, and trust in man-made engines that could fail at 300 feet?

Why would anyone chance unpredicted thunderstorms that cause long delays at microscopic airports with strange sounding names?

Certainly, there must be a more mature approach to life.

Ah, therein lies the clue. "Maturity" for the average run-of-the-mill creature is measured by such mundane items as: a secure job, a three bedroom house in the suburbs, two cars, and, if possible, some money in the bank at the end of the month. All worthwhile, perhaps, but, for the pilot, that's just not quite enough. For him, the marks of maturity go beyond the plateaus of a two-dimensional world. For him, maturity includes an unquenchable thirst to strive, to reach, to climb, to risk, and to soar to another — a vertical — dimension.

For the pilot, this vertical dimension sharply separates a "ho-hum" existence from a life lived to its fullest.

Chances? Dangers? Of course. But he wants more than tin-covered status symbols. He reaches farther to experiment with the third dimension of flight. He climbs higher to learn and obey the laws of the sky. He pushes harder to master the forces of lift and gravity, thrust and drag. He practices longer to hone his skills in his emerging new awareness.

Why would anyone want to fly an airplane into this third dimension? Maybe he wants to be mature enough to recognize the fourth dimension when it is revealed to him.

Marketing Staff
Orville
Wilbur
Designer
Engineer
Manufacturer
Mechanic
Pilot
Controler

THE WEAKEST LINK

''A chain is as strong as its weakest link.'' So runs the familiar adage. It follows, then, if every link is strong, the chain is strong.

In a sense, every profession — aviation included — is composed of links which form a chain, and advancements within each profession depend upon the strength of those links. With only one weak link, the whole chain suffers.

What is the weakest link in aviation that contributes to accidents? Opinions vary.

''It's the pilot,'' says the controller. ''He refuses to follow directions.''

''The controller is too quick with his instructions,'' retorts the pilot.

The mechanic blames the manufacturer, who blames the engineer, who blames the designer, who blames the marketing staff, who blames the Wright Brothers for starting the whole mess.

Such unwillingness to confess a wrong clouds the issue and lessens our chance of preventing a similar accident. What is really frightening is that through our lofty words mixed with the rhetorical skill of a snake-oil salesman, we might successfully shift the blame to something or someone not responsible and camouflage the cause of the accident, providing more opportunity for a repeat of the mistake.

For too long, aviation professionals have swept mistakes under the rug and have done everything possible to subcontract the heat.

Yet, our only hope for an unblemished safety record is to be honest — completely honest — with both ourselves and others. The one who refuses becomes the weak link in a chain that can save lives.

Our chain of responsibility stretches further than the distance between the pilot and the cockpit door. Pilots, controllers, mechanics, managers, engineers, flight attendants, journalists, are all links in this chain. And each one's strength is in direct proportion to his willingness to admit a mistake.

''Once upon a time, there was a king who wanted to learn how to fly.''

Parables

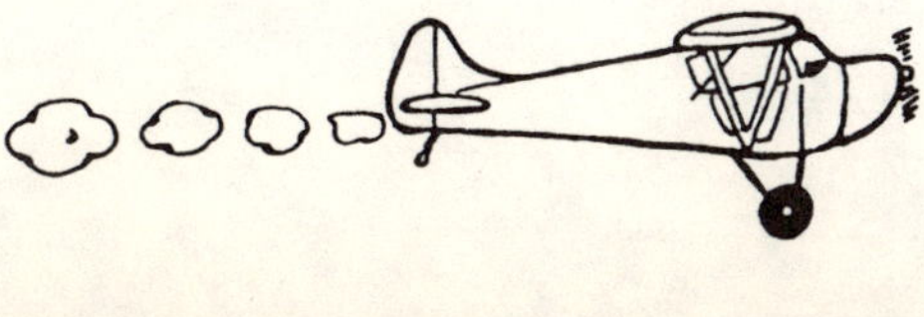

THE KING'S LESSON

Once upon a time, there was a king whose every wish was a command. To his servants he issued orders which were carried out with haste. Even members of his family responded with a sense of urgency. Wherever he went, people doffed their caps and bowed, for it was the king who had absolute power, and no one dared to disobey.

One sunny Sunday afternoon, the king announced that he wished to learn to fly an airplane, so he arranged with the local fixed-base operator of his domain to show him what must be done.

But once inside the trainer, the king felt somewhat insecure. The controls were foreign, and the dials looked ominous. As the plane rolled down the runway and became airborne, his anxieties increased. "Bah! Why should I worry?" he thought to himself. "After all, I am the king."

At 3,000 feet, the instructor turned to his new student and said: "All right, you take the controls. Try some straight and level flight."

"Straight and level," said the king. But the airplane didn't seem to understand who was at the controls. It bobbed and weaved like a porpoise. Suddenly, a sharp, buzzing sound was heard, the nose pitched forward, and the plane began to spin rapidly.

"Ye gods!" cried the king in his awkward attempts to right the craft. "This machine does not respond as I command."

The plane continued its downward spiral, until the instructor grabbed the wheel and avoided the embarrassment of an abrupt meeting of airplane with the earthly kingdom below.

"Your majesty," he said, "up here, everyone is equal. Your commands, I'm afraid, are useless. You must obey the rules."

"The rules, indeed," retorted the king; "I make my own rules."

"Not up here you don't," cautioned the instructor as he climbed back to altitude. "Here, you must learn them and obey them."

"Then I order you to make me a pilot," bellowed the king in his frustration.

"That's not one of the rules," answered his instructor. "I merely advise; everyone must learn, for himself, the mysteries of the sky."

At first, the king rebelled against this challenge to his authority; yet every time he attempted to deviate from the prescribed rules, he fell from grace and the sky, only to be rescued by the quick reactions of his instructor.

Soon, however, the king suppressed his royal ego, and yielded to the rules about which his instructor spoke. To his amazement, it was not demeaning; it was rewarding. He was not *subservient* to the rules as much as he was in *harmony* with them.

For the first time in his life, he knew what it meant to be a success. Through his own efforts, he shared the joys and thrills of peaceful co-existence with the sky.

Then, one day, without any fanfare, he removed his crown, and slipped the purple mantle from his shoulders. They were necessary no longer. His instructor was right. Up here, everyone *is* equal. Everyone is a king.

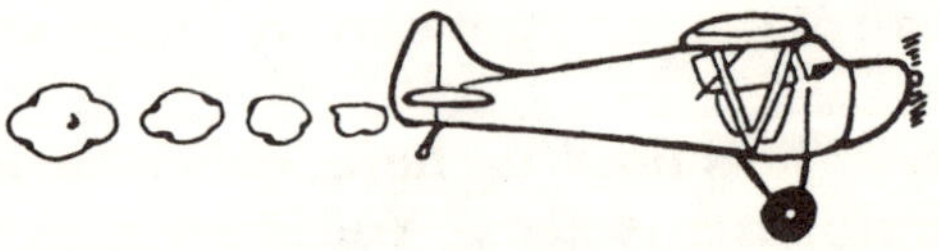

THE LIFE AND LOVES OF PETER PILOT

Ah, it was sheer ecstasy when Peter Pilot met the one he thought was the perfect girl. In Peter's eyes, she was beautiful, intelligent, and a noble creation. What seemed to be a marvelous relationship progressed from courtship to the deepest of all human emotions, love. On a blissful day in June, the two lovebirds pledged to remain faithful and to love one another as long as they both should live.

But, alas, temptation raised its ugly head. It happened on the third day of the honeymoon. While driving along a country road in southern Illinois, Peter and his bride came upon a small airfield. For reasons known only by fellow pilots, their car was attracted to the airport as if pulled by some huge, hidden magnet.

"Let's stop and take a quick look around," Peter suggested. His wife agreed.

Peter took her by the hand, leading her in a haphazard route amid the parked aircraft of assorted shapes and sizes. Suddenly, Peter stopped. His eyes rested on a weather-beaten, paint-peeled, red and white, 1951 vintage Piper Tripacer nestling near the back of the ramp. A small sign in its windshield read: "FOR SALE." It was an orphan begging for someone to give it a home. Like Pavlov's dogs, Peter began to salivate...to drool...to lust....

His bride, yet unaware of what was going on inside her husband's mind, tugged at his sleeve. "Let's go," she insisted. "There's no action around here." Reluctantly, Peter returned to the car, and the two of them continued on their journey.

As they drove, the young bride noticed a distant look in Peter's eyes. His body sat in the car next to her, but his mind remained back at that airport. She listened, patiently, as he talked incessantly about that old Tripacer. "What's gotten into him?" she asked herself.

Then, it hit her like wake turbulence from a 747. Peter was in love, again — this time with an airplane.

''Why didn't you marry that stupid airplane instead of me?'' she screamed as her patience gave way to a display of wrath unequaled in the annals of honeymoon history.

Peter remained calm, and tried to reason with her: ''Honey ...don't you understand that airplanes were made to be loved?''

''Uh...no, Peter,'' came her reply, ''I don't understand. As a matter of fact, I don't even *want* to understand.''

This revelation shook Peter Pilot to his seat belt. But he wouldn't surrender; instead, he began the monumental task of educating his new bride about the facts of life — flying wise.

Would she share his enthusiasm for flying? Only time would tell.

So, young fellows, make it easy on yourself. Never become seriously involved with a young lass, until you are convinced that she appreciates the joys of flight as much as you do. Unless, of course, you plan to find a girl with the shape of a Piper Tripacer.

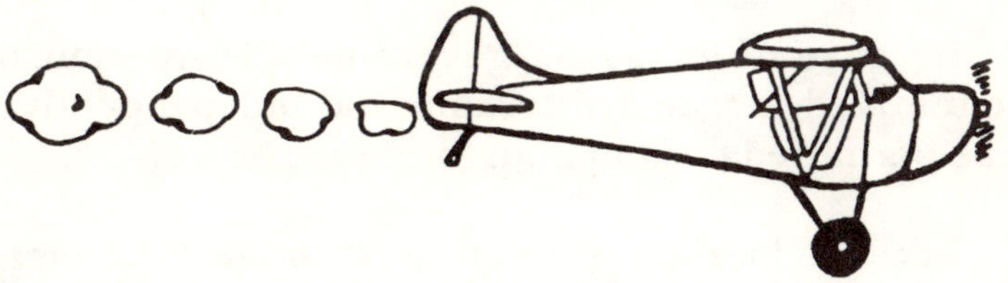

SIMPLE SOLUTIONS

We read each day in the newspapers about our nation's problems. It seems that a lot of people offer solutions, yet nothing changes. At times we raise our hands in frustration. But look how easily we could resolve these problems were we simply to encourage the government to flex its muscles and enforce strict regulations, thereby eliminating the sources of these problems.

Here are a few suggestions:

1. Because of the rapidly increasing rate of traffic accidents, only commercial buses and heavy trucks should be allowed to travel on interstate highways.

2. Because of the rising cost of education, especially at small, private colleges, only students enrolled at state-supported universities with enrollments over 20,000 would be eligible for federal scholarships or veteran's benefits.

3. Because of the growing number of alcoholics in the country, only exclusive restaurants with seating capacities of over 500 should be granted liquor licenses.

4. Because of the overcrowding at the polls and due to some poor selections of elected officials by the low income, uneducated masses, only college graduates who own at least one acre of land would be able to vote.

5. Because of increased government corruption by those who are tempted to take bribes, only the independently wealthy should be permitted to campaign for public office.

6. Because local programming often does not measure up to the professional standards of the networks, only those radio stations with 50,000 watts power would be allowed to broadcast over the airways.

And the list goes on. Each a simple solution at the price of individual rights.

Silly, isn't it. After all, these things couldn't happen today in our United States, could they?

Currently, there is a strong movement afoot to limit some major airports — supported by our tax dollars — to commercial airlines. If this happens, what would stop the others from coming about?

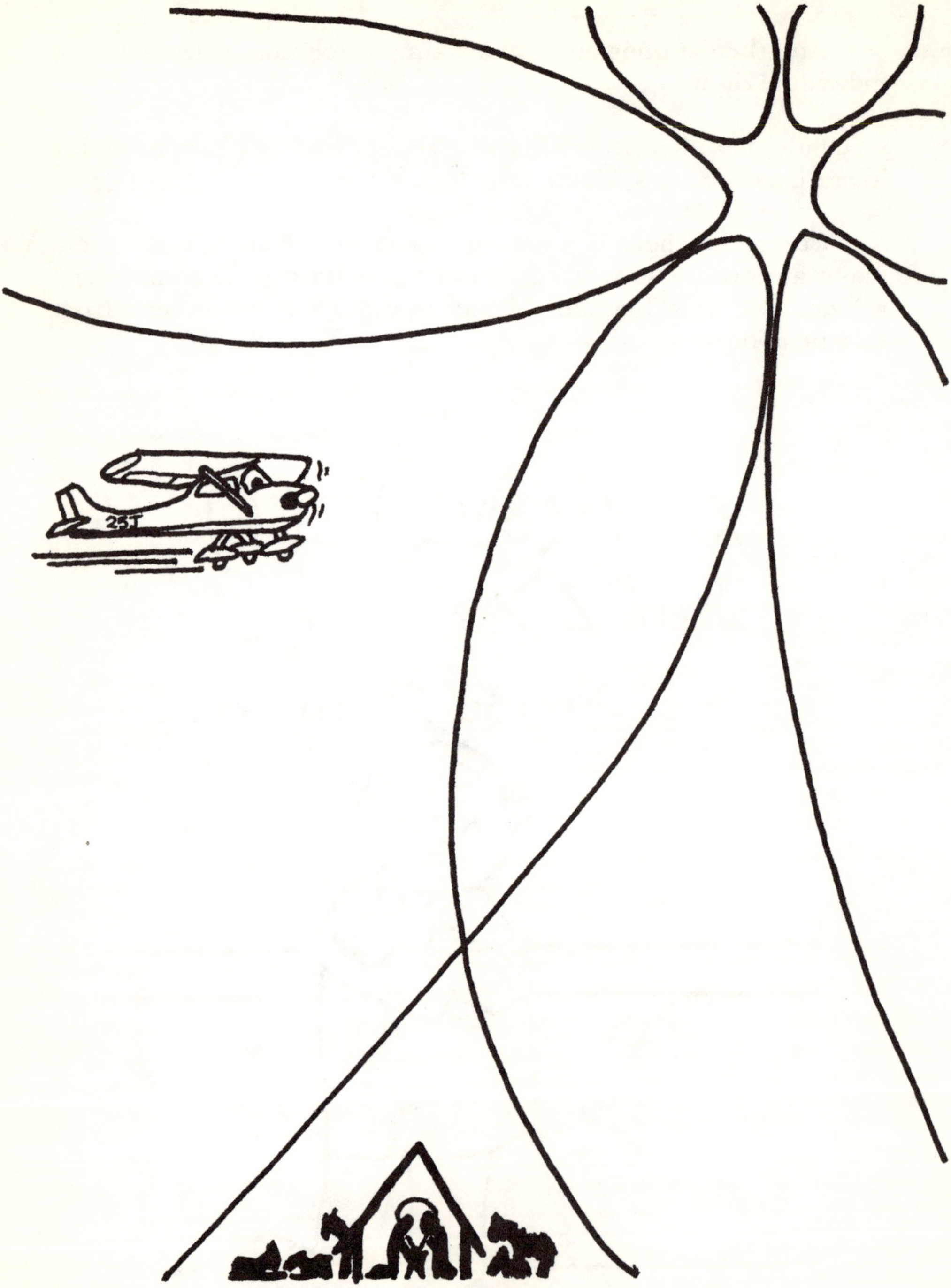
25T

PILOT'S CHOICE

So far, it had been an uneventful flight for the pilot en route to the small town airport. He looked forward to a rewarding night's sleep after his long cross-country.

Weather advisory had predicted rain for this night, but an unexpected high-pressure system moved in so that a strange calm engulfed the countryside.

The pilot passed over the capital city, and was five miles to the northeast of his destination. It was 11:34 p.m. when he called ahead for landing instructions:

25J: Uh, Bethlehem tower, this is Cessna two-five Juliett.

Tower: Two-five Juliett, this is Bethlehem tower. Go ahead, please.

25J: Bethlehem tower, I've just passed over Jerusalem intersection, inbound for landing. What is your active runway, please?

Tower: Two-five J, winds are two-niner-zero at two knots. You are cleared to runway two-seven. Report on downwind.

25J: Runway two-seven. Thank you.

Tower: Uh, Cessna two-five J, we're experiencing some interference on our radar. Please squawk 0400.

25J: Roger, 0400.

Tower: Two-five J, we show some unidentified traffic about three miles to your right moving toward the city.

25J: I don't have it in sight...uh...wait a minute...I see something...it's shining...it's really bright.

Tower: Affirmative, two-five J. We have it in sight.

25J: Wow! Look at that. It's like a...a star.

Tower: Say again, two-five J.

25J: A star. It looks like a star.

Tower: Affirmative, two-five J.

25J: It's moving. No, it's still now. It's directing all of its light downward. I can see the ground very clearly. There's a barn...and I see some men running toward it. Wow! It's wierd, I tell you.

Tower: Affirmative, two-five J.

25J: Bethlehem tower, you won't believe this.

Tower: Say again, two-five J.

25J: Above the light, I see...beings. There must be a hundred...no, a thousand of 'em. And...they're singing. Would you believe it? They're singing.

Tower: Uh, two-five J, are you declaring an emergency?

25J: Emerg...no. Man, you ought to see this. Wow!

Tower: Affirmative, two-five J.

25J: The light...it's moving away. Bethlehem tower, do you think I ought to follow it?

Tower: The choice is yours, two-five J. The choice is yours.

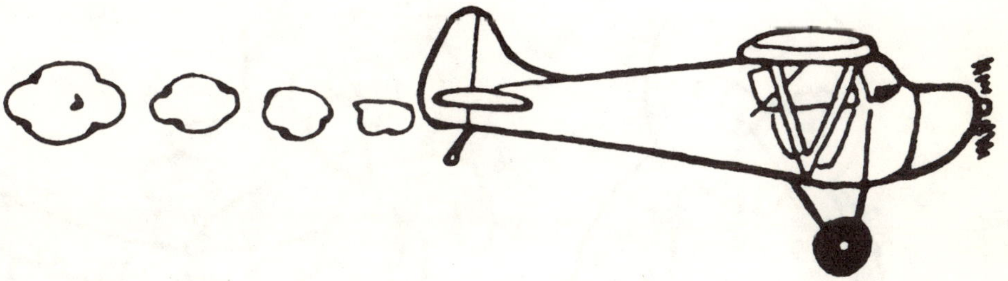

WOOSH

BURDEN OR BLESSING?

Long ago, before any measurement of time, when the earth was created, a legend tells us that God made a unilateral decision to place wings upon the backs of the birds. However, the birds, who had just learned the art of singing, raised a storm of protest, and sent a delegation to plead their cause. The delegates approached God's throne, and implored the Lord to remove their obnoxious wings.

"Is it not enough," they asked, "that you made us so light and feeble? Must we be burdened by the weight of wings?"

The Lord smiled, and, with a knowing look in his eyes, replied to the feathered delegation: "I see the justice of your complaint. However, let us wait one week, and if at that time you still desire to have your wings removed, I shall accede to your request."

Greatly pleased, the delegation left.

One day later, a strange thing happened. An unexplained force lifted them heavenward, and the birds found themselves flying and floating and enjoying a heavenly sensation.

When one week had passed, the delegation reappeared before the Lord, but this time their heads were bowed in humility.

"Merciful Father," one of them said, "how could we have doubted Your infinite wisdom? These very wings which we ridiculed as unnecessary burdens, we now cherish. They enable us to soar into the loftiest heights."

And the delegation, in unison, sang: "Amen!"

Qq
Quiet

A PILOT'S ALPHABET

A is for *Airplane* — the object of our dreams and passions. It is a combination of metal, plexiglass and rubber that, mysteriously, transforms itself into a personality.

B is for *Barnstorming* — the ultimate adventure. Here is a touch with the past which symbolizes the yearning for freedom and daring that lies within each of us.

C is for *Climb* — the pathway from the complex troubles of this mundane planet to the simple, yet majestic, wonder of the sky.

D is for *Dual* — the first phase of instruction when we learn to cope with a new dimension, to perform some basic maneuvers, and to acquire a first-hand experience with a modern magic carpet.

E is for *Engineer* — the pragmatic artist. He is the one who converts seemingly impossible ideas into workable shape and form.

F is for *Flight Plan* — without a doubt the cheapest life insurance policy in the world.

G is for *Growing* — for the pilot is always a student, constantly adding to his awareness of flight. By the way, why not begin, today, to earn that extra license?

H is for *Hangar* — the nest of our beloved birds. Scattered throughout are assorted wrenches, wheels, propellers, and an occasional exaggerated yarn that begins: "There I was, flat on my back at 3000 feet..."

I is for *Instructors* — who weaned us, guided us, scolded and praised us through hours of rigorous training. With undying patience, they molded raw recruits into qualified pilots. Instructors, bless 'em all.

J is for *Jacket* — made of dark brown leather. It comes with complimentary scarf and helmet accessories. In earlier days, each pilot was thought to own two kinds of jackets: "flying" and "straight."

K is for *Kitty Hawk* — where it all began on December 17, 1903. It is hallowed ground enshrined in the heart of every one of us.

L is for *Landing* — the acid test for the private pilot. Make a good one, and your trip is a success. Bounce once, and your passengers check the commercial airline schedules.

M is for *Mile High Club* — the membership of which probably is the result of wishful thinking more than actual experience. Frankly, there just aren't that many auto-pilots around.

N is for *Navigation* — living proof that the shortest distance between two points is a straight line, unless, of course, you come face-to-face with storm clouds, restricted areas, or a 707 at twelve o'clock.

O is for *Overanxious* — too often listed in an accident report under "pilot error."

P is for *Precision* — especially in the art of aerobatics. Relive the days of the Great Waldo Pepper. Learn to perfect an inside loop. Even the Red Baron would be proud of you.

Q is for *Quiet* — what landings, cockpits, and mother-in-laws ought to be.

R is for *Runway* — and no two look the same. Some are megamile stretches of concrete painted with numbers and fancy stripes, while others are poor excuses for cabbage patches. Yet, each is a stepping stone of flight that is the pilot's "yellow brick road."

S is for *Safe* — which aviation is, as long as you obey the rules.

T is for *Tower Operator* — the unseen voice in the sky. His rhetoric is friendly, calm, clear, and on 121.5, that of a strengthening angel.

U is for *Upward* — where every pilot is gazing, especially when stuck in bumper-to-bumper rush hour traffic.

V is for *Vertigo* — that helpless feeling when you don't know where you're going or which end is up. It's somewhat like falling in love, with even more serious consequences.

W is for *Wind* — ofttimes friendly as it glides straight down the runway at five knots, while possessing an uncanny knack for playing funny tricks in the form of turbulence or sudden shifts in direction when you least expect it.

X is for, well, X — the unknown quantity. It is the designation for all untried aircraft. It is the step beyond the drawing board awaiting some brave soul to fly its maiden voyage.

Y is for *YOU* — the one who has experienced the thrill of flight. It is you who serves as an ambassador of private aviation to your friends, neighbors, and everyone you meet.

Z is for *ZULU* — the time zone for the entire aviation community that is regulated by a zealous Pygmy at the Zulu International Airport.

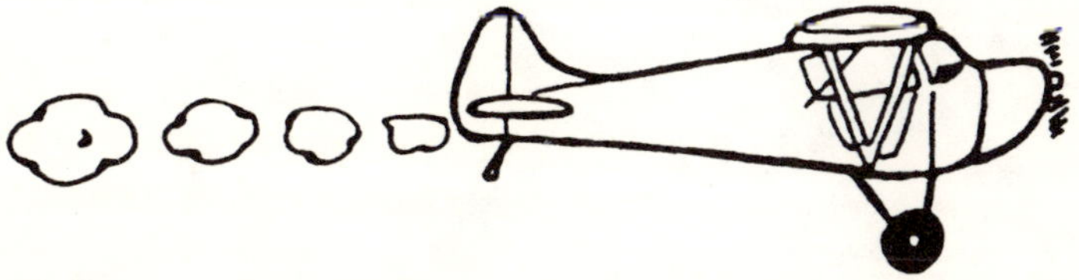

FEDERAL
AVIATION
REGULATIONS

UP HERE, TOO?

Once upon a tomorrow, a junior angel was being shown around the splendors and glories of the universe by a senior and more experienced angel. At first, the little angel marveled at the vastness of interstellar space. Soon, however, he grew tired of it all. What he saw was spectacular, of course, but it was so predictable...so precise...so mathematical....

Perhaps due to boredom, fatigue, or both, the little angel drifted from the side of his senior guide to grab hold of some floating meteor stones. He wondered just how far he could throw them in this weightless atmosphere.

"Watch out!" shouted the senior angel, as he pulled the little angel toward him. "Be sure to stay close to me at all times. Up here, you must watch where you're going."

"But why?" asked the little angel.

"The restrictions," answered his guide.

"Restrictions? I don't understand."

"Surely you've heard of the Federal Aviation Regulations part ninety-one. There are places we can't fly without proper clearance. 'Terminal Control Areas' I think they call 'em."

"Yes, ...but up here, too?" asked the young angel.

"Up here, too. Only worse. No one dares to come close to the fringes. You see, the boss can't break rules. It's not his nature."

The junior angel looked puzzled. "But isn't the boss in charge of...everything?"

"We all thought so, until the latest ruling," came the reply.

"Do you mean that the F.A.A. has jurisdiction even up here?" asked the little angel.

The senior angel gave a half-smile. "Judge for yourself," he said.

Everywhere the little angel went those next few days, he was told the same story: Several years ago, the advisory board convinced the chief angels of the Lord that the F.A.A. could be trusted. "After all," they said, "the F.A.A. consists of sound people interested in all dimensions of aviation and dedicated to protect the rights of everyone." But, alas, such was not the case. The all-protecting F.A.A., not satisfied with restricting the rights of private pilots on earth, extended its control over the domain of the angels, themselves.

"I wonder what we can do to return to the 'good old days' when angels could fly wherever they wished?" asked the now sad little angel.

But no one in heaven or on earth was able to answer.

Ah, fear not, faithful pilot. The foregoing parable is not true. The good people of the F.A.A. have not imposed their unlimited powers all the way to heaven...at least not yet.

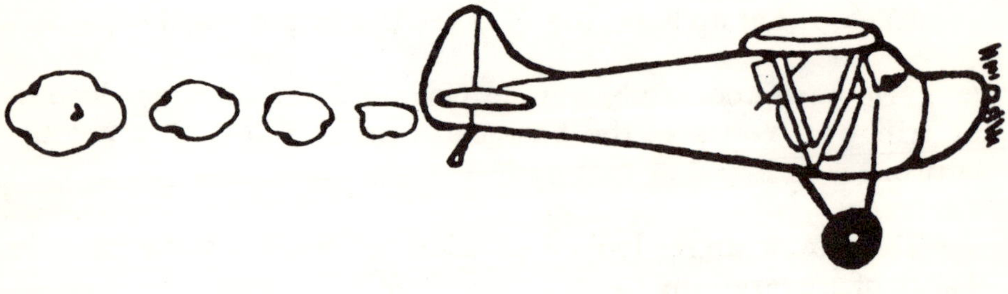

QUALIFICATIONS

Dear Mr. Orville Wright:

We regret to inform you that your application as a flight instructor at Zippo Aviation has been rejected by our personnel committee.

Frankly, we were stunned by your resume. The fact that your co-ownership of a bicycle shop heads the list as your longest held job raises questions as to the seriousness with which you regard aviation. Also, we understand by your record that your first solo flight was in a homebuilt aircraft, uncertified by any government inspector. On top of this, we learned that you soloed without any prior training in a powered aircraft. Although your initial flight lasted only 12 seconds, nevertheless, this action, alone, indicates you hold no respect for proper training.

Rumor has it that your first flight was followed by several more that same day ending with an accident (caused by a combination of gusty winds and marked inexperience) resulting in damage to your plane. We find no record of your reporting of the incident.

Zippo Aviation prides itself on its excellence in aviation training. Our instructors must comply with rigid standards. In short, your background (not to mention your apparent disregard for the rules) runs contrary to every requirement as a certified flight instructor.

Indeed, Mr. Wright, we realize that you desire to fly; this has never been in question. We humbly suggest, however, that perhaps you had best consider a career other than aviation. After all, not all persons can be successful pilots. For your own safety and that of your would-be students, we hope this personal advice will lead you to take inventory of yourself and remain in private business.

We understand that this will be a great year for bicycles.

Respectfully yours,

Zippo Aviation

Our motto: "New Ideas Cause Paper-Work"
BY THE BOOK AWARD
We're up to — HERE — in RIGID STANDARDS
PRESIDENT ZIPPO

"Somewhere above the clouds, he can touch the spirit of a lost friend."

Death

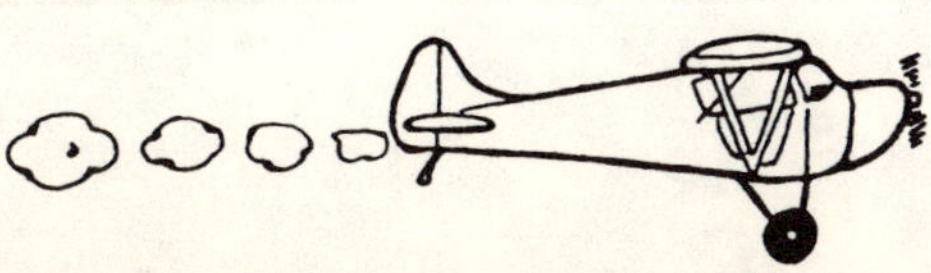

THE UNKNOWN
FAITH

SIX-ZERO MIKE

My old Aero Commander 100 is going to the junk yard. Yet she's going with dignity and glory. She's not going out with a whimper and a sigh, but with a roar and a bang.

The fellows in the junk yard can hear her coming. Her crankcase is cracked and valves stuck. She sounds more like a motorcycle as she spits and backfires. But she's traveling to her final resting place spunky and spirited.

Her body is a wreck. Banged, scraped, and nicked all over. Her once shining coat is now dull and flaky — rusting here, patched up there. Her brakes are worn, her windshield cracked. Her doors rattle in ill-fitting frames letting cold air blow through the openings. And her heater really never worked that well.

Her airspeed indicator jumps erratically during taxi, and the Hobbs meter registers nearly 7,000 hours.

By all standards she is unfit and unsafe for the sky.

But I admire Six-Zero Mike. I love her engine that turns over at the key's first nudge and the slightest crack of the throttle. I love her bounce and forward lunge despite her body's creeping and final decay. She may go down, but she'll do it unbowed and proud.

If only I can go out that way! With spirit. Alive. Bouncing. With my engine still humming. Ready, yes, for the junk yard of the grave, but awaiting a new restoration and a new awareness.

Sputtering, maybe, but not whimpering. Gunning my engine. With eagerness. Anxious. I'll roar into death, and come to a screeching halt.

And then, meet God face-to-face.

THOMAS ETHOLEN SELFRIDGE

It was more than 70 years ago — September 17, 1908, to be exact — in Ft. Myer, Virginia, that a nervous Army lieutenant, Thomas A. Selfridge, climbed aboard a five-year-old invention called a "flying machine" in order to analyze its potential to the armed service signal corps.

Nervous? Surely. But Lieutenant Selfridge gained comfort in knowing that the pilot was one of its creators — Orville Wright, himself. However, something happened that afternoon that would be remembered to this day. As the biplane neared an altitude of seventy-five feet, a loud noise came suddenly from the engine. A propeller snapped in two, and severed an overhead wire, causing the craft to tumble out of control, and fall helplessly to the earth.

Orville Wright was injured severely. Next to him lay the body of the twenty-six year old lieutenant...his broken body lifeless, where just moments before was full of zest and excitement.

History records Lieutenant Selfridge as the first person killed in an airplane crash. But neither mention in a history book, nor the naming of an airbase in Michigan in his honor, can they replace the lost life, just as they cannot erase from memory the mental anguish suffered by his family and friends.

But the consequences reached even further than his immediate circle of acquaintances. When the death of Lieutenant Selfridge or any one of the thousands of souls who suffer the same fate is reported, the entire aviation community mourns. Each pilot pauses to bow his head, and offers a silent prayer — a prayer not only for the fallen victim, but also for himself; when one member of his fraternity dies, a part of him dies as well.

Often, tears stream down crimson cheeks, and confused heads shake in utter disbelief. The age-old question: "Why?" again is unanswered — hidden in some mind greater than any on earth.

The memory of a fallen comrade weighs heavily upon his heart, yet the pilot enters his airplane again, pushes the throttle to the wall, pulls back on the stick and heads higher than he has ever gone before.

For some, this is only therapy. For others, it helps to maintain perspective. Perhaps, for a select few, it is a hope that, somewhere above those clouds, he can touch the spirit of a lost friend.

The following is a tribute by Richard J. Queenan, a close friend and admirer of Captain Charles Blair who was killed in an airplane crash on September 2, 1978:

THINK IT THROUGH

Young pilots!

As you start your career,
There's a bit of history I think
you should hear.

It's about a pilot of yesteryear
Who put his mark upon us all,
And made us glad he walked
so tall.

Charlie Blair was his name,
But though you're too young to shed a tear,
Consider this and
think it through.

He died in a plane crash
For reasons unknown
Enshrouded in a mystery left
to God, alone.

"So, who was Blair?" you might well say
"And this 'Think Through' for me today?"
Why you must know that he knew his role
A giant amongst pilots who soloed the pole.
As Pan Am captain, a million souls knew his name,
And many changed flights just to fly on his plane.

It wasn't John Wayne who married Maureen
But Blair, himself, who loved the colleen.
But through it all, one thing stood clear,
People just loved to catch Charlie's ear.
And what a smile and exhibit of grace
With superior airmanship that set the pace.

His polar fighter was a P-51,
And he made his own compass to run with the sun.
Now see Charlie's plane on Smithsonian review,
But when you look, pause and take heed.
Learn your lessons carefully,
not with speed.
For when it can happen to one of the best,
The safety legacy is passed on to
the rest.

Think it through.

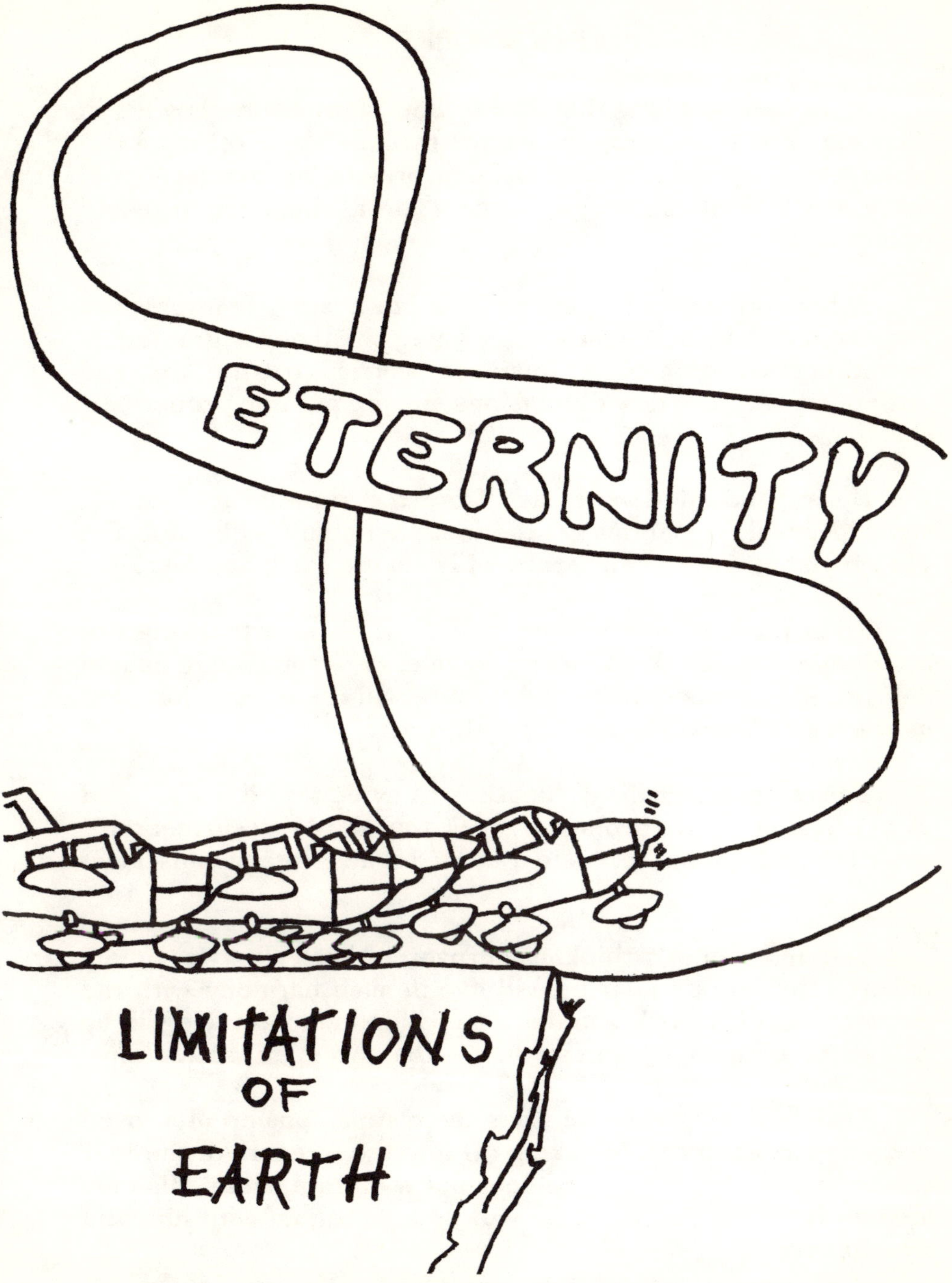
ETERNITY
LIMITATIONS OF EARTH

TAKE—OFFS

No matter how long they live or how many hours they log in the air, pilots never seem to lose the thrill of a take-off in an airplane. Any airplane — commercial or private, prop or jet — will do. Every take-off is a magic moment, promising adventure and excitement.

At take-off, a man becomes a little boy again. He forgets the humdrum activity of the past day. Things he took for granted — landscapes, clouds, people, business, worries, frustrations, and anxieties — acquire new dimensions and are put into proper perspective.

During the sudden release from the runway at take-off, some incurable romantics get the idea that man was born to fly. They feel their natural environment is not the earth but the sky.

Right now, we mortals are able to work with the forces of lift, drag and thrust in temporary segments — for as long as fuel and aircraft components last. But there will come a day for all of us when this fact will be a thing of the past.

The pragmatist calls it "death." A cynic says it is the "end of everything." A theologian uses the word "resurrection," while the philosopher refers to it as a "higher state of consciousness."

Perhaps a pilot's thinking surpasses them all. He simply believes that at this hour he will live in such harmony with the elements that his earthly limitations will be no more. He will be able to travel as far — even farther — than his mind will let him.

Realizing this, when he loses the companionship of a loved one who has ascended by taking off into the freedom of the four winds, even in his sorrow, he does not want him back. His only longing is for the day when he will be able to join with him and fly...forever.

"Let our flight plan be your will and our true course in line with your commandments."

Prayers

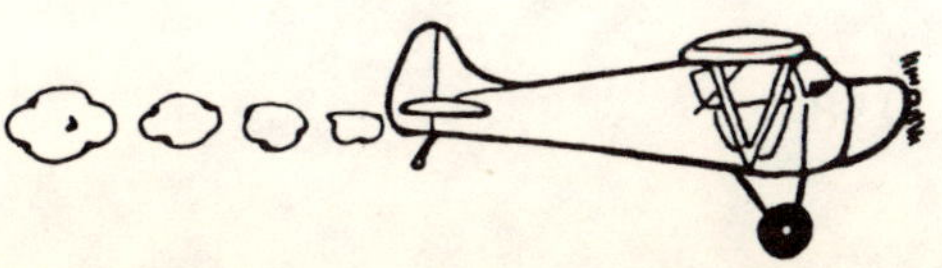

Smithsonian Institute Museum

HOLY GROUND

I went on a pilgrimage today. I was only one of thousands of Americans who, each year, rekindle their passions for aviation through a trip to the National Air and Space Museum of the Smithsonian Institute in Washington, D.C.

Each year, the displays are somewhat different, but the same spirit prevails.

I felt as though I was walking on holy ground as I stood beneath the Wright Brothers' *Flyer* made out of spruce and unbleached muslin with which it all began in 1903. I was within arm's length of Amelia Earhart's bright red Lockeed Vega 5B. I touched the capsules of "Friendship 7" and "Apollo 11." And, of course, my eyes were drawn to that Ryan silver monoplane, number N-X-211, hanging above them all. What history!

There were pictures and statues of the characters — the flesh and bone of aviation: Billy Mitchell...Baron Von Richthofen...Neil Armstrong...James Doolittle..."Pappy" Boyington ...Grover Loening.... Priceless memories.

I looked about me and saw other people react to what lay before them. A nine-year-old boy wearing a t-shirt which read "Boston Red Sox" looked quickly from left to right, unsure what to see first. A young business executive in a polyester three-piece suit tried to remain aloof by refusing to display any emotion. A middle-aged man with one leg hobbled on crutches close to a P-51 and gave a silent salute. A teenage girl fell in love with the picture of a young Charles Lindbergh.

Reactions. Everyone had his own. There it was, in one building, the fulfillment of a dream held for centuries that, one day, man would fly.

Indeed, this place is sacred. Instinctively, I bowed my head ...and prayed.

WOW...
Thanks!
YOU'RE WELCOME

THE CYCLE

One morning this week, the sun and I rose together.

At about 6:00 in the morning, while it was yet dark, I took off in a Cessna 172 from Runway 9. As I climbed higher and higher, the opaque horizon before me gradually became a distinct outline of treetops and buildings. Then, the morning sun peeked over the surface and shone in such a way that the approaching clouds distinctively reflected different colors.

First, I saw a mellow orange, then a light orange, a yellow, a bluish-white, and the texture of the clouds became more prominent. And all in the space of just a few minutes.

What a sight! The people, below, who were turning off clanging alarm clocks and turning over in bed for just five minutes of extra sleep, would miss this kaleidoscope of colorful splendor.

As I watched in wonder, I found myself forming the words out loud: "What a magnificent sunrise." Then, as a private prayer, added: "Thank you, Lord, for that beautiful sunrise."

At that moment, I realized a thing of beauty is not complete until I, or someone else, responds in praise and thanks for it.

It's a cycle, really: the giver..., the gift..., the receiver..., the thanks..., the grateful receiving of thanks....

Our praise and thanksgiving are necessary for any gift, especially for the overwhelming evidence of God's creation.

Now, that's something that every pilot ought to experience at least once in his life.

Lord,
Guard and guide
the men who
fly....

AIRMAN'S HYMN

The year was 1915. The United States was becoming more and more involved in the agonies of a global war in order to help "make the world safe for democracy."

Although it was less than twelve years since man first lifted a powered aircraft through the air, this war already was producing aviator heroes from both sides of the battle lines. Before the war's end, pilots such as Germany's Manfred von Richthofen (the "Red Baron"), Rene Fonck of France, and America's own Eddie Rickenbacker would become household names.

In this setting, a little-known poet, Mary C. D. Hamilton, captured the romance and danger that surrounded those who left the solid earth for the unchartered sky. Her verses, set to music, have been called by many "The Airman's Hymn." You might be interested in its petitions:

> Lord, guard and guide the men who fly
> Through the great spaces of the sky;
> Be with them traversing the air
> In darkened storms or sunshine fair.
>
> Thou who dost keep with tender might
> The balanced birds in all their flight,
> Thou of the tempered winds be near,
> That, having Thee, they know no fear.
>
> Aloft in solitudes of space,
> Uphold them with Thy saving grace.
> O God, protect the men who fly
> Through lonely ways beneath the sky.

Mary Hamilton never held a pilot's license, but her words keenly reflect the prayers of the entire aviation community.

Mary Hamilton, we salute you.

FLYER'S ROSARY

Max Conrad, record-setting pilot of the 1950's, died April 3, 1979 at the age of 76. According to a report in a recent issue of *General Aviation News*, Mr. Conrad logged more than 50,000 hours — nearly six years — in the air.

History will mark the records of this pioneer, and countless thousands will talk about the day they met this giant of aviation. But it was not Max Conrad's records for which I remember him. Instead, it was *his* philosophy of flight.

Like every incurable romantic, Mr. Conrad appreciated (perhaps "worshipped" is a better word) the thrill of flight each time he lifted an airplane from a runway. I doubt that he ever became bored while flying, even during his 56 hour, 26 minute non-stop flight from Casablanca, North Africa to El Paso, Texas.

Recently, I ran across a little-known expression of his faith which he entitled: "Flyer's Rosary." In Max Conrad's own words comes a testimony to the worship of flight:

By night on swift enchanted wings I fly,
Bright stars above become my Rosary,
Each star a lonely prayer which bids me try
To live in faith and hope and charity.

At times I seem to question truth above
And even doubt sublime eternity,
Yet countless stars tell me a Holy Love
Will watch and care for me, Eternally.

All through the night I prayed my Rosary
On Heavenly Beads where only Angels trod.
How can I ever doubt Life's mystery
When first at dawn my humbled soul is awed?
The generous sun gives me so tenderly
Another day that I may live for God.

A PILOT'S PRAYER

Almighty God:

At birth we were cleared for takeoff and began to practice all sorts of strange maneuvers.

Each day, as we mature and take solo flights into the dimensions of life, we encounter some unexpected turbulence.

Often, we are blown off course by the winds of hatred, prejudice, or brutal force.

Under the pressures of business, and home, and society, we experience a vertigo as to a proper sense of values.

At times, we become desperate, and are caused to shout: "Mayday," only to discover that there is no location or situation outside of your control zone.

O God, let our flight plan be your will and our true course in line with your commandments.

Then, when we make our final approach in life, may we, with all the angels in heaven, be prepared to say: "I'd rather be flying with you — today, tomorrow, and forever."

Amen.

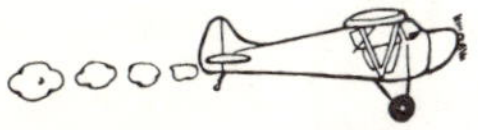

Dedicated to the memory of Dr. Tine W. Davis (1914—1980), trustee of Embry-Riddle Aeronautical University and friend of aviation.

FLYING
LESSONS
MEET HERE
NOV. 1
INSTRUCTOR

A PILOT'S THANKSGIVING

In 1863, President Abraham Lincoln proclaimed the last Thursday in November for, as he said, "a day of thanksgiving and praise to our beneficent Father." Ever since President Lincoln's proclamation, Americans, annually, pause to give thanks for their blessings of the past year. Yet, even the astute Mr. Lincoln could not be aware that, years later, this day would have special significance for a creature known as the private pilot.

Each Thanksgiving Day, as the church bells peal and the Doxologies are sung, I reflect upon the special blessings that come to all who share the joys of flight:

> For instructors and flight examiners who dare to discipline in order that I might become the best pilot possible,
>
> I am thankful.
>
> For tower operators, approach control, and flight service stations who guide me through rain, fog, clouds, and storms,
>
> I am thankful.
>
> For engineers and manufacturers who create from metal, rubber, and plexiglass a craft that assumes a personality all its own,
>
> I am thankful.
>
> For mechanics, fixed-base operators, airport managers, and aircraft dealers who care for my airplane and keep it fit for the sky.
>
> I am thankful.

For the N.T.S.B., F.A.A., and C.A.B. who speak out for me and every private pilot in this land,

I am thankful.

For wives, husbands, and children who patiently endure our enthusiasm for flying that sometimes interferes with family gatherings,

I am thankful.

For *General Aviation News, Flying Magazine, Plane and Pilot*, and all the other aviation periodicals that keep me abreast of the news and promote the cause of general aviation,

I am thankful.

For the opportunity to view America from a new perspective high above the little problems of everyday life — a dimension once reserved for the angels, alone,

I am thankful.

Finally, for my country and for my God who made all of these possible,

I am thankful — very thankful, indeed.

I think Mr. Lincoln, himself, would agree.

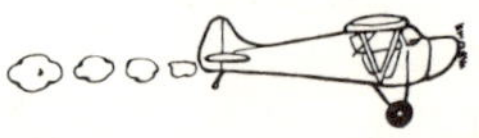